PLANT
EXTINCTION

PLANT EXTINCTION:

Distributed by
Winchester Press
220 Old New Brunswick Road
CN1332
Piscataway, N.J. 08854

A Global Crisis

Dr. Harold Koopowitz
and Hilary Kaye

Published by
Stone Wall Press, Inc.
1241 30th Street, N.W.
Washington, D.C. 20007

STATE LIBRARY OF OHIO
SEO Regional Library
Caldwell, Ohio 43724

Second Printing January, 1984

Dust jacket designed by Hasten Graphic Design
Illustrations and Cover Photographs by Dr. Harold Koopowitz

Library of Congress Cataloguing in Publication Data:
Catalog Card No. 82-062894
Koopowitz, Harold, and Kaye, Hilary
Plant Extinction: A Global Crisis

ISBN 0-913276-44-8

Copyright© 1983 by Stone Wall Press, Inc.
1241 30th Street, N.W. • Washington, D.C.

All rights reserved. No part of this book may be reproduced or transmitted in any form, or by any means, electronic or mechanical, including photocopying, recording, or by any information storage and retrieval system, without permission in writing from the Publisher, except by a reviewer who wishes to quote brief passages in connection with a review written for inclusion in a magazine, newspaper, or broadcast.

Printed in the United States of America.

86-33672

Dedication

"Grandpa, grandpa, tell me a story," the little girl begged, tossing her blond curls into the air as she hurled herself expectantly toward the old man.

He bent down, gathered her into his arms and held her high in the air over his head. She squealed with childish pleasure.

"Sure, honey, what story do you want to hear?" he asked.

"Grandpa, tell me the one about the old times!" she said excitedly.

"What about the old times?"

"Oh, you know! The old times, when there were so many different kinds of flowers that you couldn't count them all. When there were forests with hundreds of butterflies and pretty orchids and lots of big, big trees. . ."

* * *

This book is dedicated to our families, friends, and lovers with the fervent wish that the above scenario will never come to pass.

Acknowledgements

Many people contribute directly and indirectly to the completion of a book, including those individuals who suffered, in patience, during the preparation of the text, and were content to put up with both our orneriness and need for quiet. My children, Lynleigh and Michael, were able to reduce their demands for my time and I appreciate their understanding. Perhaps my greatest debt goes to those who have helped and still nuture in me a love for green plants. It is these people—Norito, Marilyn, Joanne, Charlie, Billy, and Ed—who more than anyone have encouraged my obsessions and applauded quietly in the wings. I would like to acknowledge the financial support from the Elvinia S. Slosson fund and the Institute of Museum Services that made the research into cryogenic seed storage possible.

Harold Koopowitz
October, 1982

Acknowledgements

The support, encouragement and patience of my husband, Layne Ballard, was invaluable during the preparation of this book. My daughters, Jan and Stephanie, also are due special commendations for patiently sharing me with our home computer. Friends and colleagues—particularly Linda Granell, Peggy Stern, and Helen Johnson—also provided needed moral support. We also thank Henry Wheelwright and Rachel Parker of Stone Wall Press for having faith that the layman would care about plant extinction. Last, but certainly not least, I wish to acknowledge and publicly thank my parents, Dick and Ruby Kaye, who have continuously encouraged my writing career.

Hilary Kaye
October, 1982

TABLE OF

CONTENTS

Introduction

Today's world events move very rapidly. Future predictions foresee significant technological advances, and improvements are usually expected to occur in only a few years. Unfortunately, this accelerated change is not limited to improvements. Man is not only a great inventor and builder, but he has also proved to be the most destructive force ever to appear on the face of the earth. Future predictions based on current statistics must now also include the devastation of vast areas of the remaining undeveloped surface of our world—and with that destruction, the inevitable extinction of thousands of species of plants and animals.

A great deal of attention has recently been given to attempts to reverse this pattern of destruction, or at least to preserve individual species which are recognized to be in danger of extinction. Such activities generally have been associated with birds and mammals pushed to the edge by the actions of man. Most everyone in the United States knows of the plight of the California Condor but few people are aware, that in that same state of California, more than 650 species of plants are currently endangered or threatened with extinction. In Hawaii, over 50 percent of the native plants are in danger, and the same pattern of imminent extinction is emerging everywhere.

Harold Koopowitz and Hilary Kaye have put together the story of this present threat to so many of our planet's plant species. They not only explain the nature of that threat, but they also point out the significance of this potential loss. The extinction of one more plant may not appear to be a matter of undue concern, but it certainly is. The danger to plants, their importance to man, what we are currently doing about it, and what more we might be able to accomplish in the future must concern us if we are to preserve our way of life and our future on this planet.

Dr. Gilbert S. Daniels
Immediate Past President of
The American Horticultural Society

I.

DO PLANTS REALLY MATTER?

CHAPTER 1

Who Looks Out for the Plants?

Anyone who reads the daily newspaper or watches the evening news has heard that the world appears to be teetering on the brink of an ecological catastrophe. It's common knowledge that the great whales—the largest mammals to exist—are facing manmade extinction. Most people have heard about condors and whooping cranes and the great apes. They have followed man's attempts to breed these dwindling species and rescue them from forever fading away. Conservationists have fought long and hard to save a variety of animals, from butterflies to alligators. But who looks out for the plants?

It's true that very few people are concerned about the plants. Yet the plant kingdom faces a far greater extinction crisis than that occurring in the animal kingdom. We have lost about two hundred animal species since 1500 A.D., but we lose at least twice that many plant species each year. We tend to forget that plants feed us, clothe us, and protect us from the elements. Because they are ubiquitous, we take plants for granted. They move so slowly that we consider them inanimate. Their leaves usually are not hairy enough to pet. They just sit there and look pretty. For most people, relating to a fuzzy caterpillar is far easier than relating to a plant. Many people simply find plants boring.

Nevertheless, plants are essential to human and animal life. Consider plants as the bottom layer of a set of building blocks. Without the bottom layer, all the higher levels of life would topple over. Right now the plants need spokespeople, just as animals have needed human "friends."

No one knows the exact numbers of plants that have ever existed or that currently exist. During the history of our planet, many species have naturally become extinct. The number of higher plant species has been estimated at approximately one-quarter million. This seems like a vast number, but actually there are far more animal species. On the average, eleven animal species exist for each plant species. In early 1980, scientists estimated that one or two plant species become extinct each and every day, somewhere in the world. By the end of the decade, these numbers are expected to rise to about one species every hour. By the turn of the century we anticipate the loss of between 15 and 25 percent of all higher plant species. Serious repercussions will affect all animals and all humans on earth. At our current pace, when we have lost 40,000 plant species by the year 2000, the crisis will not stop. Extinctions will accelerate until this planet has become impoverished.

Why?

What is the underlying cause of all this trouble? Very simply, the problem is too many people but not enough thinking ahead. Our world is so overpopulated that nearly 1.5 million people die from starvation each year. Three-quarters of this number are children. Despite these conditions, the worldwide population continues to grow—perhaps to seven billion by the turn of the century. The Earth probably could provide enough food for this number of people, but humans do not seem able to resolve the political problems involved with food distribution and storage. Wild plant species are wiped out when their habitats are destroyed to create housing or agricultural fields. People not only need food and shelter, but they require fuel. When they can't afford to buy coal or fuel oil, they collect and burn plants.

Where?

Plant extinctions are truly worldwide in scope, occurring from the Arctic Circle to Antarctica. The epicenter of the crisis is in the steamy rain forests of the tropics. We anticipate that all significant tropical forests will be lost by the turn of the century, except for remnants in the remote outposts of Zaire and lowland Amazonia. The rate of destruction in the tropics is almost incomprehensible. In the Amazon, about four hundred square miles of forest are stripped each year. Some islands are in terrible shape, too. Haiti, for example, was once a lush island that exported quite a few island-grown products. Today, just 8 percent of the forests are left and even sugar cane, once an abundant crop, must be imported. Since more than 40 percent of all plant species occur in the tropical zones, the loss of plant life in these regions will have serious consequences. As the human population grows, less and less land can comfortably support human life. Livable land is lost when the creation of deserts, a process called desertification, begins.

We sometimes assume that highly industrialized Western nations are relatively free of plant extinction problems. This is not the case. When the serious nature of the plant crisis first achieved widespread concern among conservationists, an immediate action was to inventory the problem. Individual plant scientists in most developed countries drew up lists of endangered and threatened plant species. The first estimates in the early 1970s suggested that about 10 percent of the species would be threatened. Now that many of the lists have been published, we see that the problem is much worse. For instance, in Hawaii more than 97 percent of the endemic species are threatened. In some areas of Eastern Europe, 45 percent of the plant species are threatened. These percentages are extreme but the overall extinction rate is now believed to be about 20 percent, twice the original estimate. This means that one in five plant species will become extinct unless something is done—and done soon. This is the critical decade. Another ten years will be too late.

Looking Back

The world has always been in a state of ecological flux. Habitats are never constant and perhaps we should not expect the earth to revert back to another Garden of Eden. We tend to forget the changing nature of the planet, probably because our own lifespans are so short compared to the hundred of millions of years of geologic time.

When the Spaniards arrived in Central America at the end of the fifteenth century, most of the land was cultivated fields with a population numbering between four and six million people. However, the combination of the Spanish pogroms and the introduction of Western diseases tremendously reduced the population. It took nearly three hundred years—until the early 1800s—for the population to climb to 1.4 million. With this smaller number of people, the forests regrew and covered the land. Today, more than 18 million people live on the isthmus between the two American continents. Now the forests are nearly gone again. Forests are resources for people, but what do the people do when they run out of resources? Looking to El Salvador and Nicaragua, we see two countries without significant forests. Political turmoil already exists there and will grow worse as the ever-burgeoning populations must deal with less and less material security. In Nicaragua, deforestation has occurred so quickly that the primary lumber tree, *Macrohassaltia,* is on the verge of extinction. It was not even scientifically classified before it was obliterated.

Looking to the Mediterranean, few people remember that the stony hills of Spain and Greece were once covered with forests. These were chopped down long ago to feed the medieval fireplaces of Europe. Even the great Amazon was not always forest. During the Pleistocene (more than two million years ago) the world endured a very dry climatic period. The Amazon forests were reduced to seven small patches or *refugias*, as they were known. When the climate changed, the forests spread and covered much of South America.

If the forests come and go, why are we worrying about them this time? The reason is that this particular deforestation is so thorough that there is little chance of the forests ever being resurrected. In the past, sufficient remnants were left to permit the forests to regenerate. This time we are like a swarm of locusts, devouring everything in our paths.

Students frequently ask if new plant species evolve fast enough to replace those lost. The answer is no. We are obliterating species far, far faster than they can evolve. Evolution takes thousands and sometimes millions of years. This destruction, in some cases, is occurring in decades.

A Few Terms

Throughout the natural world the individuals that compose a species are usually counted in millions. Sometimes a greater number of individuals die than are born or germinated. This initial decline in numbers may signal the beginning of the extinction process. If the trend continues, a critical stage is reached where the population collapses. Without remedial action, the species slides into extinction. Conservationists use a number of terms to describe species in different stages of this process.

Animal and plant species in any stage of the extinction process are listed in Red Data Books, which are volumes published by conservationists at the national and international levels. The primary publisher of these books is the International Union for the Conservation of Nature and Natural Resources (IUCN) in Switzerland. The IUCN has defined Red Data categories which specify the exact stage of the extinction process. Not all countries use the same terms or even the same definitions for the terms. Since the majority of countries do use the following terms, these are the ones we have used in this book.

VULNERABLE species are the ones whose population numbers are decreasing and are likely to become more severely threatened with time. *ENDANGERED* species have populations that have become so reduced that they are in eminent danger of becoming extinct. *RARE* species have a population of less than 20,000 individuals. Some species are naturally rare and have never occurred in greater numbers; yet they are able to maintain those numbers. Other species become rare through the actions of man or other unnatural forces. Unless measures are taken, these species will become vulnerable. Organisms no longer known to exist in the wild are considered to be *EXTINCT*. The IUCN definition includes species that are extinct in the wild but survive in cultivation. We prefer to modify the terms so that *EXTINCT* means entirely absent and *EXTINCT IN THE WILD* refers to species that are still alive in captivity. *THREATENED* is a broader term that is used for species that fit into any of the above categories. These terms and definitions will apply throughout the book.

Is There Any Hope?

What can be done? Actually, relatively little can be done when we consider the extent of the damage. It is beyond man's present capabilities to turn the planet back to its original pristine condition, as many conservationists strive to do. What we can do is to build a technological ark to save as many species as possible for the future.

The most obvious solution to the problem is to set aside conservation preserves where the remaining threatened species can be protected. However, to be effective preserves must be carefully managed. This requires an intimate knowledge of the area's ecology and a lot of money. Unfortunately, most of the world (and especially the Third World) is critically short of money.

While we hope to see as many effective preserves as possible, we believe that a combination of modern technology and effort can create a so-called "ark" that can save threatened species for future generations. We're referring to the creation of cryogenic gene banks. Most kinds of plant seeds and spores can be processed so that they can be frozen for thousands of years to be thawed and germinated in the future. A few banks are already in existence but the technique has hardly been utilized. Gene banks can be set up with minimal expense and expertise, as we will explain later in this book.

The Effect on the Shambaa

In the Western world we tend to think of plants primarily as resources for food or ornaments in the landscape. Some people recognize the many products produced with forest woods. Sadly enough, so many of us live so far from the earth that we either do not know or have forgotten the origins of a dining room table, a cotton dress, a drugstore prescription, or a sheet of paper. The products pass through so many hands along the way that the final consumer usually has little concept of the original source of the product.

People live closer to the earth in other parts of the world, and their dependence on plants is recognized on a daily basis. The inhabitants of these less-developed areas are forced to cope constantly with the problem of diminishing wild species. We see one example of this dependency in a study done by Anne Fleuret, an ethnobotanist, who studies the interactions between man and plants.

Fleuret has analyzed ways in which the Shambaa people of northern Tanzania use plants for purposes other than food. While these people have a less-developed lifestyle than that found in Western countries, they do use machine-made products. However, their daily lifestyle involves the use of many "local" products, either in raw form or that have already been processed into purchased goods.

The Shambaa live in the Usambara Mountains at high altitudes, either in small villages or in clusters of dwellings. As late as the 1940s, the extensive forests of the area appeared to be untouched. Today, only remnants are left and there is considerable pressure to convert these small patches into agricultural lands.

One of the biggest demands for forest products is for building houses. The walls of the Shambaa houses are traditionally constructed with poles, through which more flexible branches are interwoven. Mud or plaster is then applied to the framework. In the local forests, a tree fern grew that had stems up to 8 feet tall. These stems were particularly resistant to decay and insect damage, and lasted over thirty years. Most other poles were good for only five to ten years. The preferred tree ferns (*Cyathea usambarensis*) now are scarce; the houses today are considerably less durable. The hardwoods of the forests are used to make door and window frames, as well as the doors and window shutters themselves. Unfortunately, the numbers of mature trees that can be used for construction have dropped. Timber cutters now need licenses which are difficult to obtain. Furniture, too, is made out of the same hardwoods. Consequently furniture items are now both more scarce and more expensive.

The availability and quality of little, everyday items also has been affected by deforestation. Today, items such as wooden spoons and the handles of tools must be made with softer woods from plantations, not the preferred hard woods of the forests. Much of the food eaten by the Shambaa people involves preparation with a mortar and pestle. The mortars are purchased but the pestles are personal objects collected by women. The best pestles are made from branches of the Dombeya tree. These trees are quickly disappearing and the search for pestles is now an arduous task. Many other household items made from forest products are also now more difficult to obtain. The list includes baskets, woven sleeping mats, storage containers, tooth cleaners, toilet paper, gum, dyes, and toys.

The greatest demands on the forest are for fuel. Fleuret estimates that the yearly wood consumption for fuels is 3,500 pounds per person per year, or, put another way, fifty pounds of wood per household per day. She surveyed three hamlets with a total population of 175 people and worked out that their fuel needs required 1,360 mature acacia trees per year. Deforestation has now reached a point in this area that women responsible for gathering the fuel spend up to 50 percent of their day searching for and carrying fuel. In some areas everything that burns is collected, including bark, roots and even dried leaves.

As this example shows, people in developing areas of the world are feeling the plant extinction crisis sooner. Those of us living in developed Western nations will feel the crisis later. The only way the impact can be lessened is if people become more familiar with the problem, learn why plants are essential to our lives and then take action—even a small action—to save what we still have left. Reading this book may well be the first step.

Disa uniflora

Case History: The Flower of the Gods

At the southernmost tip of Africa lies Cape Town, the pearl of the Southern Seas. The city is perched on a strip of land that juts out at the exact point where the Indian Ocean meets the Atlantic Ocean. Its beaches boast the whitest sand and the bluest waters, while the rear of the city is guarded by the citadel of a mountain, a flat-topped peak aptly called Table Mountain. It looks as if God had knocked off the tip of the mountain with a swipe of His hand. Each afternoon, a cloth of cloud coagulates atop the table and sends moisture dripping over the sides. *Disa uniflora,* one of the most fabulous orchids in the world, has chosen Table Mountain for its home. These orchids usually are found lining the banks of the small streamlets that run down the mountain.

Disa uniflora has a small rosette of nondescript, strap-shaped leaves gently nestled among sedges or moss. Despite this humble beginning, the plant thrusts upward a central shaft bearing four, five, or even more large hooded blossoms that range in color from cerise to scarlet. The orchids can measure up to six inches in diameter but most plants have somewhat smaller flowers. The natives have named this spectacular orchid the "Flower of the Gods."

The species used to be quite common. People walking along the streets of Cape Town frequently would see great bunches of these flowers hawked at roadsides. Even as late as the 1960s, blooms could be purchased at florist shops for just fifty cents apiece. Where are the red Disas now?

At the turn of the century, huge batches of Disa plants were exported to England and Europe. Descriptions left behind from those years tell us that these flowers were easy to grow and were treated like tender, tuberous Begonias. In fact, these beautiful orchids once were grown as bedding plants. They were sheltered during the winter and then planted in the late spring to enjoy the drizzly English climate. The Great War, however, brought their casual popularity to an end. As English gardeners rushed off to the trenches to be killed, their knowledge of how to cultivate Disas went with them. The few individual plants that survived the war eventually disappeared.

After World War I, the South African government introduced legislation to protect those Disas left in the wild. This halted the wholesale plundering of these plants for Europe, but flower picking and small-scale pilfering continued. Like all wildflower legislation anywhere in the world, the laws were mostly ineffective. Disas, despite these well-intentioned laws, continued to diminish. Attempts to reestablish Disas in cultivation appeared doomed and the plants quickly earned a new reputation for being difficult to cultivate. Seed from Disas turned out to be equally stubborn.

Swedish gardeners in Gottingen finally discovered a way to first germinate and then grow the scarlet Disa. At Kirstenbosch, the national garden of South

Africa, successful cultivation also has been achieved. Although other botanical gardens have been less successful, amateur growers in South Africa now have perfected ways of growing seed. The popularity of these Disas is starting to rise once more, but we do not yet know if this will be enough to save them. A few plants in one or two collections and a rapidly dwindling wild stock does not bode well for the long-term survival of the Flower of the Gods.

CHAPTER 2

Medicinal Plants: Poisons that Heal

Frequently we're asked, "Why worry? What does it matter if one more plant species becomes extinct?" The answers are numerous and will be explored in detail in this and subsequent chapters. One of the most important reasons for saving plants is the tremendous medicinal potential stored within many species. Plants are wonderful chemists, a trait that benefits not only the plants themselves but also humans. Chemicals inside the plants create a powerful self-defense system that fends off vegetarian animal predators. These plants have evolved chemicals that poison their animal foes by interfering with the animal's own biology. Although the chemicals sometimes are deadly to the animal, these same products found in the leaves, roots, bark and flowers can be useful medicines for people when the doses are designed for the human body. We will look at some examples of plant chemistry in this chapter.

The Historical Perspective

In today's urbanized world, the green serenity of botanic gardens is a soothing tonic for those often surrounded by concrete and asphalt. Our ancestors also enjoyed the beauty of botanic gardens but in those years the gardens mainly served to soothe and heal. Botanical gardens were developed and maintained by medical schools in past centuries. Botany was not a separate academic subject but rather a sub-discipline of medicine.

Before synthetic chemicals dominated medicine, as they do today, roughly 80 percent of all drugs were derived from plant materials. Chemists eventually developed synthetic versions of many plant drugs, but these manmade products never would have existed without nature leading the way. Even today, when manmade chemicals such as plastic and polyester reign supreme, at least one-fourth of all medicines still are prepared from plant material. In some cases, chemists have not yet learned to duplicate nature. In others, the synthetic version, dependent upon petroleum byproducts, is too costly.

People have recognized the medicinal value of plants for thousands of years. Even though our earliest ancestors may not have understood how or why certain plants cured specific ailments, they were well-aware that plants heal as well as nourish. Written records of medicinal plants date back nearly 5,000 years to when both the Chinese and Sumerians recorded use of hundreds of medicinal plants. Modern scientists have been unsuccessful in their attempts to duplicate the healing effects of some of these plants; however, tests with other species show that our ancestors' prescriptions often were right on target. Drugs sitting in our medicine cabinets today actually are age-old remedies.

Whether they are old or new remedies, medicinal plants are big business. Roughly 25 percent of medical prescriptions come from higher plants, totalling about $6 billion in annual sales. Altogether, about 40 percent of the prescriptions come from natural sources: 25 percent from higher plants, 13 percent from lower or microbial plants and 3 percent from animals.

The Pill

A medicinal plant that ranks near the top of the list of important plants is *Dioscorea,* a yam that grows wild in the jungles of Mexico and other tropical regions. Nearly one hundred different kinds of *Dioscorea* grow in Mexico alone. Most of these yams have no known special value but a handful of species have dramatically altered the medical world. Compounds called "saponins," which allow chemists to produce semi-synthetic hormones, are found in the roots of these species. These hormones, also called steroids, are included in oral contraceptives and cortisone drugs used to treat a variety of illnesses ranging from arthritis to ulcerative colitis to skin disorders.

Cortisone's value was partially realized even before the discovery of *Dioscorea.* Cortisone, when it first appeared on the scene, seemed to be a new wonder drug that could ease the crippling, painful symptoms of rheumatoid arthritis. This new medicine, however, could be produced only when a natural "starting material," a hormone, was used in the synthetic process. In those pre-*Dioscorea* years, cortisone drugs were developed with a starting material extracted from the bile or urine of animals with adrenal glands, such as cattle, sheep, and hogs. Not only was this process incredibly expensive, but it involved a difficult, lengthy procedure. The demand for cortisone drugs far exceeded the very small quantity available.

Even though the belief that cortisone cured rheumatoid arthritis later proved to be overly optimistic, the premature idea did spark an international search for a cheaper, easier way to produce cortisone. The search soon turned away from animals to focus instead on plants, which were known to contain sterols. Since sterols are the plant equivalent of the hormones present in humans and animals, it seemed logical that they, too, could be used as a starting material. *Strophanthus,* an African genus, was the first group to be seriously tested. However, further investigation proved that it had little medicinal value. Too few sterols were contained within the plant's seeds. Nevertheless, scientists had proven that cortisone could be produced with plant sterols—they just had to find a plant with a sufficient amount.

The sterol-rich *Dioscorea* was finally discovered during the 1940s by Dr. Russell Marker, an American organic chemist who was an authority on plants containing sterols. Marker's familiarity with these species led him to the Mexican yam. He tried to interest several large pharmaceutical companies in his plan to travel to Mexico to further investigate *Dioscorea* species but the companies doubted his ability to pinpoint a practical source of sterols and refused to finance his research. Undaunted, Marker traveled to Mexico on his own, set up a primitive laboratory in Mexico City, and finally uncovered a handful of *Dioscorea* species that contained saponins in substantial quantities. He called the saponin found in these species "diosgenin," the same compound used today to manufacture synthetic cortisone. Millions of people today use cortisone in one form or another to relieve pains associated with everything from football injuries to rheumatoid arthritis.

The most astounding attribute of *Dioscorea* was uncovered in 1956, when Dr. Gregory Pincus created an effective, easy-to-take birth control pill using diosgenin as the starting material. An earlier birth control chemical derived from steroids had been produced before *Dioscorea* came on the scene, but it had to be injected. This made it an unlikely choice for most people. The contraceptive derived from *Dioscorea* is taken orally, a feature that sets it apart from all other contraceptives we know of today.

Surrounded by the high technology of the 1980s, it is easy to forget that if it were not for the nondescript Mexican yam, we might have neither birth control pills nor widespread cortisone drugs.

The Heart Regulator

The number one killer in the United States is still heart disease. In 1980, the National Center for Health Statistics reported that each day an estimated

3,400 Americans suffer heart attacks and about 1,600 others suffer strokes. Deaths due to heart ailments have decreased during the past few decades but the problem still is severe. Imagine what the problem would be like without *Digitalis,* a plant commonly called foxglove. Roughly three million Americans treat their heart ailments daily with medicine derived from the delicate foxglove plant. The secret to this plant's success in treating heart disease lies in the glycosides found within its leaves. These substances regulate a faulty heart by strengthening the contractions and permitting rest between beats.

Patients who suffer from mild to moderate heart conditions find relief with a preparation of crushed *Digitalis* leaves. Those with more critical heart problems require a more potent medicine. These stronger remedies contain one or more of the individual glycosides found within the leaves, usually digitoxin, gititoxin or gitoxin. *Digitalis* is not the only plant to contain glycosides, but it is considered the best all-around source. Oleanders, lilies of the valley, wallflowers, and others also contain these compounds in varying amounts. Patients undergoing actual heart failure usually are prescribed Ouabain, a drug with glycosides even more effective than those in *Digitalis.* Derived from *Strophanthus gratus,* the potent Ouabain must be injected and therefore is used only when quick action is needed. *Digitalis* is preferred as a daily heart regulator because it is taken orally and is less toxic.

When the British physician William Withering discovered in 1775 that a powder made with foxglove leaves relieved the symptoms of dropsy, *Digitalis* began its impact in the medical world. Dropsy was a disease of unknown origin or cure. All that doctors knew was that the primary symptom, swelling, resulted from a mysterious accumulation of liquid in the body. Withering tested *Digitalis purpurea* in response to folk tales told in the countryside of England and discovered that it was a diuretic. By reducing liquid in the body, the drug "cured" dropsy. Articles written by Withering show that he suspected that the foxglove medication affected the heart, but he did not realize that dropsy was only a symptom of heart disease.

Nearly one hundred years later, a French chemist isolated the glycoside digitonin. Soon after, doctors finally realized they were treating a heart ailment when they prescribed *Digitalis.* Knowing this, they were able to treat heart disease before the dropsy symptoms occurred. The more potent medicines used today were developed through the isolation of the three glycosides within *Digitalis.*

Of the millions of patients who have been prescribed *Digitalis*, about 20 percent experience side effects that range from nausea to convulsions. Since we already know of other species that contain cardiac glycosides, perhaps there is another species, still undiscovered, with abundant cardiac glycosides that would cause fewer side effects. If such a species exists, we can only hope that it has not yet fallen victim to extinction or is not among the currently endangered species that probably will die out before the turn of the century.

An Old Indian Cure

Hypertension or high blood pressure, an illness that goes hand-in-hand with heart disease, also is treated with a plant drug. Even though hypertension is a rather painless disease, it can be deadly since hypertensive people are more prone to heart attacks and strokes. Their constricted blood vessels force their hearts to work harder while pumping blood through their bodies. For most people with hypertension, the best remedy is the plant drug called reserpine.

Reserpine, an alkaloid found in the roots of *Rauwolfia,* has a dramatic effect on the body's sympathetic nervous system. The drug relaxes the blood vessels which, in turn, reduces the heart's workload. The discovery of reserpine as a hypertension treatment was accidental and followed its initial use as a sedative for mental patients, particularly those with violent, schizophrenic behavior. The observation that mental patients had lower blood pressure following treatment with reserpine prompted further tests with hypertensive patients.

Rauwolfia, a shrub native to India, was used by Indians for thousands of years to treat snakebites, scorpion stings and some forms of mental illness. Western science, however, ignored these treatments in India until the last part of the nineteenth century when two Dutch scientists reported success with *Rauwolfia serpentina.* Even then, forty years passed before scientists and physicians really took notice of the Indian shrub. Clinical trials with mental patients began and plant hunters started investigating a variety of *Rauwolfia* species—not only the Near Eastern plants but also species found in Central America and northern South America.

Powdered *Rauwolfia* was somewhat effective; yet its effects were erratic. In 1952, chemists finally isolated the alkaloid responsible for the medicinal effects. That discovery enabled the drug to become highly effective. Chemists called this compound reserpine and used it exclusively to calm down mental patients. During this period of widespread use during the 1950s, doctors recognized the side effect of reduced blood pressure. Today reserpine is mostly used to treat hypertensive patients.

Chemists have learned to synthesize reserpine, but the process is more expensive than simply extracting the alkaloid from the plant. Most pharmaceutical companies, therefore, still rely on the natural product.

Malaria Relief in the Tropics

Just as heart disease is the number one killer today, malaria used to be mankind's biggest threat in earlier centuries. Malaria caused more deaths throughout history than did all the wars and plagues combined. It remained mysteriously untreatable until the 1630s when a group of Jesuit missionaries living among Peruvian Indians stumbled upon a cure. The missionaries observed that a drink made from the cinnamon-colored bark of a native Peruvian tree seemed to ease the dreaded symptoms of malaria. Whether the missionaries or the natives discovered the magical bark's powers, and how, are still unclear. The missionaries did not know how or why the cure worked, but they happily reported the find in Europe.

The Peruvian natives called the tree "quina" or "quina-quina"—from which the name quinine was drawn—but an often told, though false, tale led to calling the tree Cinchona. According to the story, the wife of the Count of Cinchona was desperately ill with malaria and was cured when she drank the tonic made from the Peruvian tree. Thrilled with her own good health, the countess gave the cure to the poor people in Spain. The tale may be false, but the tree still bears the erroneous name.

News of the miraculous malaria cure spread through Europe. The conservative medical community of the seventeenth century regarded the new medicine as quackery, but the bitter brew worked so well that soon it became very popular. Until the turn of the nineteenth century, no one knew exactly how much Cinchona bark cure was needed to bring about a cure and dosages varied. All that changed, however, when two French chemists isolated quinine, the active alkaloid within the bark. This allowed doctors to prescribe precise doses of uniform potency. The discovery of quinine also halted the need to drink the noxious brew. Instead, patients could pop quinine tablets into their mouths and achieve even better results. After the initial discovery of quinine, chemists isolated nearly forty different alkaloids within Cinchona. Most of these appear to have no particular medicinal value; yet one additional alkaloid, quinidine, successfully treats certain heart diseases.

Chemists produced the first synthetic quinine tablet shortly before World War II. This early drug was hardly competition for the natural product since synthetic quinine cost about $1,000 per gram and caused a variety of unpleasant side effects. A few years later, chemists created another synthetic version. This one compared favorably with natural quinine and grew to be a popular substitute. The synthetic product was used frequently during World War II, partly because the abundant Cinchona plantations in Java were cut off to the Allied troops during much of the war. The malaria-infested troops fighting in tropical regions quickly turned to the new synthetic product.

Before World War II, the wealthy owners of the lush plantations in Java held the Cinchona monopoly. The synthetic anti-malarial soon caused the monopoly to crumble. After the war, many people were certain that the natural product was no longer needed. We now know that this willingness to discard natural quinine was premature. In certain parts of the world—mostly South American and Southeast Asia—certain strains of the protozoa that cause malaria (via mosquito bites) can no longer be contained with synthetic quinine. Natural quinine, however, still is highly effective against these strains. Even though temperate areas today have little trouble with malaria, the illness should not be considered simply a page in history. People in many tropical regions of the world still dread this deadly disease. Enough people fortunately recognized the need to preserve Cinchona, and these trees still grow in abundance on plantations.

As we see with Cinchona, synthetic replacements are not always perfect substitutes for nature. What happened with this anti-malarial could happen with any synthetic medicine. When it comes down to choosing which species to save, all known medicinal plants—whether or not they have spawned synthetic versions—should be high on the list.

Cinchona also illustrates how natural genetic variation within individual plants and closely related species can be used to create more productive plants. The first commercial Cinchona trees contained about 4 percent quinine. In those years before plantations were created, Cinchona bark was randomly collected and bark with the highest yield of quinine was used. As farmers and scientists became aware of Cinchona's tremendous economic value, they carefully bred trees that were exceptionally rich in quinine. The quinine yield in Cinchona eventually jumped from 4 to 13 percent.

A Serendipitous Discovery

When doctors in Canada and the United States first began examining the medical potential of periwinkles, they were unaware that they were hot on the trail of a cancer treatment. Because folk doctors in several parts of the world were known to use periwinkle leaves to treat diabetes patients, research teams in both countries independently began testing whether or not periwinkles affected blood sugar levels. Despite the encouraging folk tales, neither group found a link between diabetes and periwinkles, properly called *Catharanthus roseus* (or *Vinca rosea,* its earlier name).

Both groups, however, discovered that alkaloids within periwinkles are anti-cancer drugs. The Canadian group found that a drug made from periwinkles reduced the white blood cell count in laboratory animals without significantly affecting the animals in any other way. The doctors recognized the drug's potential to counter an elevated white blood cell count (a key factor in certain forms of cancer).

By 1958, the Canadian doctors had isolated vinblastine, the alkaloid that reduced white blood cells. Vinblastine was found throughout the periwinkles but the leaves were the best source because they contained the highest concentration and permitted the easiest extraction.

While the Canadian team was checking out the folk medicine tale from Jamaica, an American doctor was studying periwinkle species in his Indianapolis laboratory. He likewise did not anticipate finding an anti-cancer drug and, ironically, his team also was seeking an anti-diabetes drug. His exploration was due to his recollections that soldiers stationed in the Philippines during World War II used periwinkle leaves when they were unable to obtain insulin.

Despite the amazing similarity of their pursuits, neither the Americans nor the Canadians knew of the other's work until the Canadians presented their findings in 1958 at a conference in New York. The Canadian group isolated the new anti-cancer alkaloid and the American team developed the technology to produce vinblastine in large quantities. The Americans also isolated a second alkaloid, called vincristine, with anti-cancer properties.

Vinblastine and vincristine both are presently used in cancer treatments; vinblastine is most useful for patients with Hodgkins Disease and vincristine is the preferred treatment for children with acute leukemia. Before the discovery of vincristine, children with leukemia had a 20 percent chance of survival. Now the odds have jumped to 80 percent. Neither drug is considered a complete cancer cure and neither eliminates the usual side effects associated with chemotherapy. Nevertheless, many patients have shown improvement with vinblastine and vincristine, either alone or in combination with other drugs.

Catharanthus roseus originally comes from Madagascar, an area where plant extinction rates are horrible and growing worse. We cannot help but wonder if a periwinkle relative verging on extinction, or one that has already been lost, might not have contained a chemical with even more effective anti-cancer properties.

Take Two and Call Me in the Morning

While most synthetic drugs are derived from natural plant compounds, other drugs such as the ever-popular aspirin follow an indirect route. Aspirin is the most popular remedy today. The little white tablets are taken for countless reasons, the most frequent being pain and fever reduction.

The history of aspirin—its proper name is acetylsalicylic acid—goes back to the ancient Greeks who used extracts from the white willow, *Salix alba,* to ease pain. No one knows for sure how the Greek physician Dioscorides learned of the soothing powers of willow juice, but we know that for thousands of years natural products from the willow were applied externally as an antidote for pain. North American Indians also reduced pain and fever with the willow.

During the early part of the nineteenth century, a French chemist isolated a compound that appeared to be the active medical ingredient in the willow. He called the compound salicin, derived from *Salix,* the genus name for willows. Salicin is contained in other species as well, such as *Spiraea ulmaria,* or queen of the meadow. Salicin effectively curbed pain but had one serious drawback: it could only be used externally. Internal doses caused severe gastrointestinal symptoms. A decade later, chemists learned to synthesize the natural drug salacin. They called the new pain remedy salicyclic acid but it, too, could not be tolerated internally. Instead, salicin was used in external treatments such as dandruff shampoos and foot powders for athlete's foot.

Finally, in 1899 German chemists created a drug that could be swallowed, the medicine currently known as aspirin. Aspirin is not produced from any part of the willow, nor is it a substitute for any natural plant compound. Yet the pain remedy is patterned after salicyclic acid, the synthetic substitute for the salicins found in willows and queen of the meadows.

From the Andes to the Dentist's Office

A drug commonly associated with the dentist's office came into existence earlier this century only because of cocaine, a natural drug derived from coca leaves growing wild and in cultivation in the South American Andes. The dentist's drug that numbs our gums and cheeks is procaine, more commonly known by its trademark name, Novocaine. The anesthetic properties of Novocaine are vital to both dentists and eye surgeons.

People have chewed coca leaves for thousands of years to stimulate their nervous system and depress hunger. Coca, or *Erythroxylum coca,* originally was considered the property of the royal Inca family but later was chewed throughout South America. Despite its popularity, no one understood how or why coca affected the body as it did. It wasn't until 1860 that German chemists isolated cocaine, the active ingredient of coca.

The drug's medical life began in 1884 when a doctor working in the Viennese clinic of Dr. Sigmund Freud discovered that a tiny bit of cocaine dropped into the eye permitted painless eye surgery. This initial discovery led to cocaine being used as a local anesthetic in a variety of minor operations, although the drug did create other problems. Because cocaine is extremely toxic, doctors were forced to use small doses. Safe dilutions often meant that the drug's numbing effect did not last long enough. Furthermore, doctors were anxious to avoid the stimulation that accompanied the numbing effect.

Scientists began searching for a synthetic drug that would be a longer-lasting anesthetic without also being a stimulant. The search ended with the creation of two synthetic drugs, procaine and xylocaine, both of which are widely used today.

The Helpful Fungus

A widely used remedy that is both a childbirth aid and a migraine headache treatment comes from ergot, a fungus that contaminates grains. The fact that ergot, or *Claviceps purpurea,* can play a role in childbirth has been known since medieval times when European midwives fed ergot-infested rye bread to women in labor to speed up the delivery and reduce the loss of blood during the procedure.

During the Middle Ages ergot caused the disease known as St. Anthony's Fire or Holy Fire. Thousands of people died from this illness, which contracted blood vessels and resulted in arms and legs becoming grotesquely gangrenous. The disease acquired its unusual name because the medieval people believed that a pilgrimage to St. Anthony's Shrine would cure the illness. The journey sometimes worked, but not because of spiritual powers emanating from the shrine. The pilgrimage sometimes was successful because the person traveled away from home and frequently changed his regular diet on the journey. The traveler's new, rye-less diet sometimes halted the symptoms if they were not too advanced. St. Anthony's Fire is hardly known today, mostly because people now know its origin and take better care in milling rye flour.

Claviceps purpurea is cultivated today because of its two medical uses. The original use by the medieval midwives is still valid today, although women in labor no longer need to eat contaminated rye bread to be helped. Women now take ergonovine, an alkaloid found in ergot, or methylergovine, a semi-synthetic derivative. Both drugs stimulate uterine contractions and decrease postpartum bleeding. Another ergot alkaloid, called ergotamine, is used to treat the severe pain that comes with migraine headaches. This drug counteracts the dilation of arteries and arterioles in the brain that occurs during these attacks.

The Leafless Wonder

Ephedra, an unusual-looking low shrub with green twigs and no leaves, is another plant that was recognized as a healer thousands of years ago. This strange conebearing species grows in many areas of the world, including China, Russia, the American Southwest, and Mexico. Ephedra's history goes back to 2700 B.C., when the Chinese emperor Shen Nung described various illnesses that could be cured with *Ephedra.* The Chinese brewed an infusion of *Ephedra* twigs and used this tea to increase blood pressure, reduce fever and suppress coughs.

The Western world did not exploit the plant's medicinal powers until the last part of the nineteenth century, when pioneers in Utah and Indians in Mexico treated syphilis with a tea made from *Ephedra.* Doctors today do not believe *Ephedra* cures syphilis, but they realize that alkaloids found within the twigs are valuable in treating a variety of other illnesses. During the 1880s, *Ephedra's* active ingredient, ephedrine, was isolated by Japanese and German scientists. Ephedrine is widely used today in the United States and elsewhere in treatments for bronchial asthma, hay fever and low blood pressure, particularly the decline resulting from spinal anesthesia. Since ephedrine was synthesized in 1927, both the natural product and the synthetic version have been used. The natural product originally was extracted only from *Ephedra* plants in China. When trade ceased between China and the West, doctors realized the need to obtain the natural product elsewhere. *Ephedra* plants naturally growing in the American Southwest lacked medicinal properties, but suitable alternatives to the Chinese species were found both in India and Pakistan. More recently, the species native to China, *Ephedra sinaca,* has been cultivated successfully in the southwestern states of this country.

The Potato's Helpful Cousins

Plants within the potato family, Solanaceae, have spawned a large group of drugs that contain what are called the "belladonna alkaloids." The main three belladonna alkaloids—atropine, hyoscyamine, and scopolamine—are used widely to treat illnesses ranging from Parkinson's Disease to hay fever. Solanaceae plants richest in belladonna alkaloids include *Datura stramonium,* or jimsonweed; *Atropa belladonna,* or deadly nightshade; *Pituri duboisia,* and *Hyoscyamus niger* or henbane.

These extremely potent alkaloids are being prescribed in small doses. Belladonna alkaloids are popular in ophthalmology, where they aid diagnoses by dilating pupils. Doctors also use these medicines when patients have abdominal muscle spasms, asthma, Parkinson's Disease, and hay fever. One alkaloid, scopolamine, is the most effective remedy for motion sickness. These alkaloids have been synthesized but the laboratory versions are more expensive than the accessible, garden varieties.

The Remedy for Gout

A popular plant enjoyed by both plant hobbyists and doctors is *Colchicum autumnale* or autumn crocus. This plant yields colchicine, one of the first remedies found to wield a very specific action on the body. While most other plant drugs can treat several different symptoms and ailments, this drug is useful only as a cure for gout, a painful inflammation that usually begins in the big toe. Colchicine has no effect on pain from any other type of inflammation, such as from arthritis, which makes it a good diagnostic tool for doctors. Colchicine is extremely toxic and can cause additional problems if the dose is too high.

Tomorrow's Drugs

With so many plants providing one medicine or another, or at least leading the way toward synthesis in the laboratory, it might seem as if all of the valuable medicinal plants have already been discovered. That's hardly the case. Less than 5 percent of all plant species have been analyzed as potential medicines. Potential new plant drugs drawn from the 95 percent of the plants still to be analyzed are being studied in laboratories around the world. As long as plants and diseases both continue to exist, new species will be scrutinized in the search for new cures. Modern technology may also help scientists find new medicinal compounds that previously went undetected. Valuable medicinal plants probably are growing right before our eyes but we simply do not understand what they have to offer.

Let's look at a potential new medicinal source—the group of plants called *Oenothera* or evening primroses. Studies indicate that these plants may strengthen our arsenal against disease. Dr. Peter Raven, well-known botanist and director of the Missouri Botanical Gardens, has spent more than twenty years studying this group of plants. He reports that some of these primrose species contain in their seed an oil rich in gamma-linolenic acid (GLA), a polyunsaturated fat and essential fatty acid. The other abundant natural source of GLA is human milk.

Fatty acids are important because they insure proper functioning of membranes in the body. Recent studies show that modern human populations have certain deficiencies in the essential fatty acids. These deficiencies appear to lead to a number of common diseases, including eczema, arthritis, and diseases of the arteries. GLA appears to be the single most important fatty acid needed to treat these diseases.

Before this medical link was uncovered in the late 1970s, primroses were viewed as nothing more than beautiful wildflowers that opened during the evening hours. Altogether, about 120 primrose species exist throughout the world, including about sixty species in the United States. While many species grow abundantly, four species in the United States are considered to be endangered, a fifth species is considered threatened, and a sixth, little-known species, native to California, should also be considered endangered, according to Raven. If GLA were found in equal amounts in each of the 120 primrose species, we wouldn't have to worry about the species being in trouble, at least not from a medical standpoint. However, as we have seen with nearly all of the previously discovered medicinal plants, the amount of the active ingredient varies tremendously from one species to another. We still don't know which of the 120 species ultimately will prove to be the best source of GLA. We may discover that one of the endangered species is the only practical source of this fatty acid. If so, efforts must be taken to prevent the species' extinction. Considering these primroses, it is important that even obscure wildflowers be preserved.

Looking Ahead

Not all of the remaining species can be saved. But instead of preserving species on a totally random basis, a priority list should be established. First, relatives of all known medical plants should be saved in the hope that one of the related species might house compounds even more effective and/or less discomforting than the original drug. One of these wild species may be just what the doctor ordered. Second, it would be wise to look into the future and try to predict medical trends for future generations. We should seek out plants that are the most likely candidates to combat the predicted diseases of the future. Of course there is no way to make accurate predictions without a crystal ball, but some assumptions can be made. For instance, it is easy to see that more and more people will be bothered by stress-related illnesses created by situations such as crowding, pollution, and toxic substances in the workplace. A likely source for medicines able to ease the symptoms of combined psychological-physiological illnesses could be plants containing hallucinogenic substances. Modifications of these natural compounds may prove to be very effective. A little forethought by today's generation may go a long way toward filling the medicine cabinets of tomorrow.

28

Dracaena draco

Case History: Dragon Trees

In a place called Tagororo on the Canary Islands, a sacred tree grew that bled a blood-red sap when wounded. The apothecaries called it Dragon's Blood. On another island, on the other side of the African continent, a sister tree grew that also bled. The collected sap of the second tree, also blood-red, was called cinnabar. The sap from both trees, when burned, was believed to be a potent defense against evil magic.

The sacred tree from the Canary Islands was *Dracaena draco* and the sister species was *Dracaena cinnabari* from the island of Socotra. They were commonly referred to in tandem as the Dragon Trees.

Of all the individual Dragon Trees, one particular specimen achieved special distinction. The Great Dragon Tree of Orotavo was known from the fifteenth century until it was destroyed by a hurricane in the nineteenth century. This was one of the most famous trees of its time and was believed to be roughly 6,000 years old. Not only was it considered to be the oldest creature on the face of the earth, but it was thought to be the sole survivor of the great biblical flood. In all likelihood the tree probably was nowhere near 6,000 years old, but certainly it was the oldest monocotyledonous plant ever known. The trunk, eighteen feet in diameter, had a small chapel built into the hollowed area where the local priest and his small congregation celebrated mass.

Dragon's Blood had a number of interesting uses. Natives of the islands used the sap to embalm their dead. Europeans used the sticky, red substance in a variety of other ways, ranging from a treatment for gonorrhea and other diseases to a varnish for staining the wood of Italian violins.

The Dragon Trees have a thick trunk that forks in different directions as it reaches toward the sky. At the tip of each branch a tuft of sword-shaped leaves juts out. Greenish flowers are borne in loose clusters and the cherry-shaped fruit is bright orange-red. The fruit is said to be slightly bitter but edible. Early reports of the fruit described it as having a small dragon on each berry, but these tales were merely the result of fanciful imaginations. The trees grow easily from seed but the seeds are very vulnerable to grazing animals. Once they begin to germinate, however, the seeds only take about thirty years to grow into mature, flowering trees. The species from the Canary Islands has been brought into cultivation and is a common landscape plant in subtropical regions such as Southern California.

Originally both Dragon Trees were plentiful. These days wild populations of both species have dwindled. *D. draco* is now extinct on four of the seven Canary Islands and no more than a couple hundred trees exist on the remaining three islands. Two islands, Madeira and Gran Canary, each have just five of the trees left. Information about *D. cinnabari* is scarce. We do know that its native island, Socotra, is in a critical situation with three-quarters of its endemic species in trouble. A relative of these trees, the Nubian Dragon Tree of North East Africa, is similarly threatened.

In earlier times, Dragon's Blood was daubed on doorposts to protect the household from ravaging enemies. What a pity these unusual trees are unable to protect themselves.

CHAPTER 3

Feeding the World

How many people can the world support? At the beginning of the twentieth century, about one billion people inhabited the earth. Today the number has risen to more than four billion people. By the year 2000, between eight and nine billion people—more than double the present number—will inhabit the planet. Starvation already is one of the greatest causes of misery and death. How will we feed everyone? How did we get into this dreadful situation in the first place? More importantly, is there a way out?

For more than a century scientists have known that in a closed system the number of animals produced is determined by the available food and space. Typically, the numbers of organisms grow in an exponential fashion until limited by either lack of food or space. If food is the limiting factor, mass death occurs. When space is limited, animals develop abnormal patterns of behavior and plants grow into small, spindly specimens. A population initially expands and then starts to climb exponentially. A small plateau is eventually reached where the birth rate equals the death rate for a short period of time. The death rate finally starts to exceed the birth rate, causing the population to plummet. This is how nature controls populations in a closed system.

For all intents and purposes, the planet Earth is a closed system and human populations are in the exponential growth phase of the curve. Since we are biological organisms, we will follow the fate of other organisms in a closed system unless we do something about it.

The solution sounds simple: stabilize births and increase food production. Unfortunately, birth rates are exponential and food increases tend to be geometric. While it would be more important to stabilize births, all indications are that this will not occur, at least not on a worldwide basis. In that case, how can we make the most out of food production?

Agricultural scientists tell us that it takes about ten pounds of fodder to make one pound of flesh. In other words, ten pounds of hay are needed to support one pound of beef. This is why meat is getting more and more expensive and, consequently, why it has become a less important part of our diet over the last century. Many more people can be fed with produce and cereal than with meat. Nevertheless, meat contains certain essential amino acids in greater abundance than do most of our present crop proteins.

Of the quarter-million plant species, about 3,000 species have been used as food. Of these, only two hundred species have been domesticated and are cultivated for food. Just thirty species are considered important food crops. The top four crops are sugar cane, wheat, rice, and maize, in that order. All are grasses. Since most of the harvested sugar cane consists of water that is lost in processing, wheat, rice and maize are truly the crops that feed the world. In 1977, just fourteen different crops yielded more than fifty million metric tons, and half of those were grasses. More than 70 percent of all farmland is planted with cereals. These crops provide mankind with more than half of its calories. It is unfortunate that most of the cereals also are low in the essential amino acids.

The critical problem we face is not how to grow enough food to feed the present world, but rather how to produce enough to feed the future world when its population has doubled, tripled or quadrupled. We now grow enough to feed the present world but about half of the crops are lost to pests and vermin. We also have trouble efficiently distributing the food to the hungry. Many people have pointed out that the problems are not so much agricultural as they are related to distribution. Immense losses occur after harvest in tropical countries. Even if these distribution problems could be corrected, the improvement still would not counterbalance future population growth. A particularly thorny problem complicates the situation. New cities are needed to house the increased population. These cities are often built on good, arable land. For example, the wonderful farms of Orange County in Southern California—farms that can give two, three, or even four crops of vegetables each year—are being paved over to support skyscrapers and apartment complexes. In Egypt, the population of Cairo is growing so fast that new, arable land produced via water from the Aswan Dam can scarcely compensate for the farmland submerged by the spreading city.

In many of the developing nations, population growth puts increased strain on a country's ability to feed its people. Yet the problems of some countries seem small in comparison to others. In 1975, Mexico used one arable hectare (two and one-half acres) to feed two people; by the year 2000, this same land must be able to support five people. This will indeed be a problem for Mexico, but consider the situation in South Korea where in 1975 the country already had fourteen people being supported per arable hectare. In less than twenty years that number will rise to 22. Is it possible for South Korea to feed all its people? If not, who will?

If we must feed people using only plants, we certainly can't do it with just cereals. We will have to use legumes—the edible seeds of crops like beans, peas and soya beans—in order to supply the correct amino acids the body needs to make proteins. A reasonable diet, in terms of these amino acids, would be two parts cereal and one part legumes. The trend in recent years has been to drop legume production in favor of cereal. Just one ton of legumes is produced for every ten tons of cereal in Asia. In India, while harvests of cereals increase, the actual yield of legumes has decreased.

The Green Revolution

During the last thirty years or so, great strides have been made in increasing crop productivity. Scientists have engineered new strains of wheat and rice with tremendously increased yields. These improvements have led to the hope that world starvation might be averted. In Punjab, India, prior to 1966, the wheat yield from one hectare was 0.6 tons. Now, from the same land, new, improved strains of wheat yield an average twenty tons. Food production there has tripled without having to bring more land into cultivation. India now is self-sufficient in terms of both wheat and rice. Nevertheless, the country may have to import grain if there is a particularly severe or prolonged drought. The story is similar for rice. Yields went from three tons to five tons per hectare. The Philippines, which used to import rice, now can provide enough rice to fill its own needs. The country can even export the grain occasionally. Even Bangladesh is approaching self-sufficiency in cereals.

All of this sounds almost too good to be true; in fact it is too good to be true. These high-yield strains are incredibly expensive because they are energy intensive, which means that they need nitrogen-rich fertilizers. These crops also require herbicides and pesticides produced with petroleum products. Besides this expense, most of the high-yielding strains are genetically uniform. This means that if a disease or pest were suddenly to appear, the entire crop could be wiped out. Such catastrophies have occurred throughout history. The great Potato Blight in Ireland is one famous example where dependence on a single uniform crop led to mass starvation. Farmers recall the recent Southern Corn Leaf Blight, a disease that reduced the United States corn crop by 15 percent in 1970. In some states the loss was even greater—as high as 50 percent of the potential crop. With population pressures on the rise, the world cannot support many occurrences of this type. The number of people in developing nations who die following even a 10 percent drop in crop productivity is staggering. Most people associate Sri Lanka (Ceylon) with tea, but Ceylon used to be a major coffee-growing country. When the fungus disease Coffee Rust appeared in Ceylon, the entire coffee crop was wiped out because the plants had no resistance to the disease. The farmers were ruined and the economy plunged into near-disaster. The situation stabilized once the island switched over to tea production. In recent years Coffee Rust has been discovered in the Brazil coffee fields. Now the same economic collapse could possibly ruin coffee farmers in Brazil, too.

If uniform crops are so risky, why have they become the standard among modern crops? The answer lies in the fact that crop machines have replaced manpower in modern agriculture. To make it possible for machines to harvest crops efficiently, all of the individuals within the crop must ripen at once and be of the same size and shape. To achieve this uniformity, the plants must be genetically identical clones or, at the very least, genetically similar. Unfortunately, uniform plants are equally susceptible to disease. Although many crops today are uniform, hundreds and sometimes even thousands of other varieties have been bred in past years. Even with some garden variety vegetables, plant breeders introduce a few hundred new ones each year. With all of these varieties already in cultivation, why should we bother to save wild species for food plants? A few illustrations will show that wild species can play an important, if unpredictable, role in agriculture.

Tales from the Vineyards

Wild grape (*Vitis*) species once were abundant in both North America and Eurasia. The Europeans domesticated the only native species to produce table and wine grapes. At the same time, the North American Indians and the early settlers merely allowed the thirty American species to ramble wild. According to legend, when the early Vikings discovered America, they called it "Vineland," an apt name when you consider the profusion of vines twisting through the New England forests and other East Coast trees.

Like many other fruits, grapes do not breed true from seed. Individual grape varieties are propagated by cuttings or by grafting onto a root stock. During the latter half of the nineteenth century, a catastrophe occurred in the vineyards of France. Insects that attacked and destroyed the roots of grape vines suddenly appeared in unprecedented numbers. Farmers could do nothing but watch as vineyard after vineyard was destroyed. The European wine industry collapsed. For a while, it appeared that the industry was obliterated. However, while the insects hungrily ravished the roots of European grape species, they did not like some of the North American species. Probably either the roots of these species did not secrete the chemical that attracted these insects or else the plants produced a chemical that repelled the pests. At any rate, grapes grafted onto wild American stock proved to be safe. Gradually the vineyards were replanted on safe American root stock and once again the grapes flourished.

The next time you drink a toast with fancy French champagne or enjoy a good Napoleonic brandy, also toast the wild American species that make these drinks possible.

Salty Tomatoes

While hindsight is perfect, there's no way to predict when a wild species will be needed. One example is what is happening with the tomato. Wild tomato species are natives of Central and South America. Some of these species were taken to Europe by the early visitors to those countries and the plants began to be grown as ornamentals in that region. Because tomato leaves are unpleasantly pungent and the plant was recognized as belonging to the deadly nightshade family, the early Europeans assumed that the fruit of these plants was poisonous. For years, no one thought to eat the plant's fruit. Only in the twentieth century have tomatoes risen to the prominence they now enjoy.

Tomatoes are one of the world's largest crops with literally thousands of different kinds already cultivated. Plant breeders have produced everything from giant 18-foot tall plants with large fruit to dwarf plants bearing tiny, round berries. There are yellow, pink and even white varieties of tomato to go along with the usual red fruit we see. Perhaps the strangest tomato was one variety that is now lost. The fruit from the variety Pruder's Purple was the color of a purple-black eggplant and weighed in at a hefty five pounds. Another strange variety had peach-shaped fruit with fuzzy skin. The tomato has proven to be very adaptable, easily converting into a myriad of varieties. With all of these unusual varieties already known, why bother to save wild tomato species? Surely by now we must possess an immensely broad genetic base for this crop.

The truth is, saving genes is a game of chance since no one knows when or which genes will be needed. For example, one of the problems facing late twentieth-century agriculture is the deterioration of water quality, particularly in those arid regions where only brackish water is available or where evaporation creates high salt concentrations in the soil. On the Galapagos Islands, a wild tomato species, *Lycopersicon cheesmanii,* grows on sand dunes where it is wafted by the ocean's salt spray. The fruit is inedible—a miserable, bitter little berry. Years ago, few, if any, people would have suspected that this species might become valuable in the future. Yet this wild tomato, with its bitter inedible fruit, has genes that permit it to survive under salty conditions. Plant breeders at the University of California, Davis, have cleverly used these genes to breed salt water tolerance into tomato plants bearing edible fruit. These plants can be watered with brackish water. There is a very real possibility that these wild tomatoes will play an important role in future farming. Who would have guessed fifty years ago that the wild Galapagos tomato would be useful?

Other parts of tomato plants besides the fruit are useful to mankind. The pungent leaves are reported to repel insects, and recent studies suggest that the leaves may also contain a fungicide effective in combating athlete's foot and other human fungal skin conditions. Because we cannot predict the unusual uses for individual species, no one knows what wondrous uses are lost when a species goes extinct.

Help from Goat Grasses

The history of wheat is long and checkered. Wheat is the most popular cereal, mostly because the flour made from wheat contains the protein gluten. When moist, dough that contains gluten becomes sticky and can be worked into bread. The first wild wheats appear to have been harvested in the Middle East. The grain probably was initially not sown, but merely collected from wild plants. The first cultivated wheats had hulled seeds with brittle heads. Alongside the cultivated fields were weedy goat grasses which were accidentally hybridized with the cultivated varieties, not just once but twice. For all practical purposes, these grasses were rather useless by themselves; however, they did contribute two important sets of genes to wheat. One set increased the gluten content of the grain and other gave the new wheats greater tolerance to environmental extremes. This second improvement was vital because the old wheats could not adapt to other climates and therefore could not be spread out of the Middle East. The history of wheat hybridization has not stopped. Plant breeders today are constantly on the lookout for new and better features to incorporate into their stock. The many wild, untapped species may hold traits that will make wheat an even more useful crop.

Some plants have special roots where bacteria are encouraged to grow. The microorganisms are able to take nitrogen from the air and convert it into nitrates and nitrites, both of which are needed by the plants to prosper. This process the plants conduct is called "fixing" nitrogen. If cereals also had these nitrogen-fixing root nodules, they would not need expensive nitrogen-rich fertilizers. Scientists for years have tried without success to create hybrids between cereals and other plants with nodules. Some South American grasses recently were discovered that are able to fix nitrogen. These species have not yet been used in hybrids but they provide hope that plants could be created that are not dependent on costly fertilizers. Perhaps there are other grasses with these nodules that will prove to be even better in hybridization.

Growing wheat in arid regions is a problem. In Australia, however, an annual grass grows from seed to maturity on a single watering. Part of the species' ability is due to its exceptionally rapid growth. If the correct genes from this grass could be incorporated into a cereal, farming then could be extended into very marginal areas and the number of crops per year on good land could be increased.

One of the most exciting discoveries in recent times has been the perennial Teosinte in Mexico. This wild relative of corn lives for many years. These genes could possibly help create a tropical corn plant that needs to be sown only once, but then would provide harvests for many years.

Sometimes wild weeds can contribute in very unexpected ways. In some rice paddies there is a tiny, wild water fern called *Azolla* that floats on the water surface. People noticed that when the fern was present it stimulated the growth of rice. It turned out that the fern had the equivalent of root nodules. Living symbiotically with the fern is a blue-green algae that can fix atmospheric nitrogen. The nitrogen is released into the water of the paddy and then is absorbed by the rice, thus cutting down the need to apply fertilizers. These ferns are now being encouraged to grow in certain regions.

The Potential of Genetic Engineering

Until now, when we have talked about transferring or using useful genes we have meant using traditional plant breeding methods. Looming on the horizon is the ability to extract the genes we need from one organism and insert them into another organism. This already has been done to a limited extent by placing animal genes into bacteria and then using the microorganism to make large quantities of products such as insulin, interferon or human growth hormones. How realistic is it to hope that we can do more complex transfers? For the foreseeable future we are going to need to rely on traditional methods of plant breeding. Genetic scientists today understand plant genes and their controls at the molecular level far less than animal genes. They still need to be able to recognize the particular genes that need to be transferred; then they need an effective way of transferring the genes. But the process does not end there. Genes must be turned on and off at the correct moment during development. There are regulatory genes and mechanisms which are poorly understood. To take an extreme example, the products made by a tomato plant in its fruit are very different from those made in the plant's roots or leaves. The plant could not function if it tried to photosynthesize with its roots or if it tried to make red, fleshy leaves that ripen. Individual plants and animals exist with great orderliness. This order does not occur by chance. We do not yet understand very well how this precise regulation occurs, but there is great potential for understanding these processes.

Soybean seeds are filled with four storage proteins that contain all the amino acids that animals need in their food. The genes that code for the protein are turned on during early seed formation and then turned off as the seed matures. The genes are never expressed in the mature plant; thus only the soybeans contain protein. If the gene repression could be eliminated and the entire plant could contain protein, we could harvest the leaves directly for protein. The tonnage of vegetable protein would be many-fold greater than that now produced from the beans. Or, to follow another scenario, perhaps the protein genes could be transferred to a bacterium that, in culture, would directly produce proteins. The result? Instant tofu or bean curd.

We eventually expect genetic engineers to be able to accomplish these amazing feats. But will the geneticists still have the genes they need to transfer by the time they are ready to carry out this engineering?

New Foods

We have domesticated relatively few of the vast numbers of plants species. Truly new vegetables emerge rarely and slowly. Nevertheless, the potential for new crops remains. Although many crops have been perfected in ways our ancestors could not have foreseen, no new major crops have been domesticated since recorded history. Many of the major crops were introduced when Europeans discovered America, but the records of plants such as maize, squash, chili peppers, peanuts, beans, and potatoes, among others, show that they had been brought into cultivation thousands of years before the birth of Christ.

While many wild species could be converted into new crops, agricultural experts believe it is unlikely that significant new crops will be developed. The main stumbling blocks are the economics involved with developing a new crop, the difficult task of educating people to accept a new food and the problem of introducing a new crop to widespread agriculturalists. In spite of these difficulties, one old and forgotten crop may eventually be resurrected. Amaranth, or pokeweed, was cultivated in the Americas roughly 4,000 years before the Roman Empire. This plant has a high protein content in its green leaves, and its seed heads can be thrashed to produce a kind of grain with a very much higher protein content than cereals. Plant breeders have scoured Mexico for variants and are now hoping to breed varieties with larger seed heads that are amenable to the threshing machines of modern agriculture. Success by the breeders and the ultimate acceptance of this crop remains to be seen.

Nearly any plant with large, underground storage organs can be bred to produce a new food. This is the story of how pretty flowers subverted one crop and saved another. In 1789, a small packet of seed was sent from Mexico to Spain. The seeds grew into flowers with swollen tuberous roots that are members of the daisy family. The plant was named after one of the leading botanists of the day, Andreas Dahl. Dahl believed that the tuberous roots of the dahlia would make a good substitute for the potato. Although for a time the plant became accepted as a vegetable in France, in the end it disappeared from favor. Right from the start, however, the species was a beauty; plants showed variation both in flower form and color. The still-popular dahlia was retained for its beautiful flowers—not for its food value.

Ironically, the potato owes its initial acceptance to its flowers. The potato was introduced as a potential food source into Europe but did not find acceptance. Perhaps people recognized that the plant belonged to the deadly nightshade family and suspected a poisonous affinity, just as they had at first suspected the tomato. The French court set upon a scheme to make the plant desirable. A royal potato plot was cultivated and the flowers were worn by the nobility. Wearing potato flowers became a status symbol. Guards were set around the potato plot during the day, but because the field was unguarded at night, plants were stolen and grown for their rather miserable-looking blossoms. Finally, after many years potatoes won acceptable as a food source and became the staple diet in many countries. People have historically been reluctant to change diets and accept new major crops. New green vegetables, on the other hand, are easier to introduce into diets. Too bad that vegetables usually are not dietary staples.

Saving Seeds

As wild species of plants have perished,some varieties of garden vegetables available to the public have also disappeared. We now are losing many different garden varieties, partly due to seed companies. Large chemical companies are buying up small seed companies and, since they are interested in maximizing their profits, many of the companies no longer offer seed varieties that are not popular. This has tremendously decreased the number of vegetable varieties available for the backyard gardener. With the arrival of improved cereals and grains, many agriculturalists expected native strains of cereals to be replaced and therefore lost. Rather surprisingly, this has not turned out to be the case. Because the importance of genetic variability in crops was recognized early on, many significant gene banks have been set up to conserve the seed of agricultural crops.

Dionaea muscipula

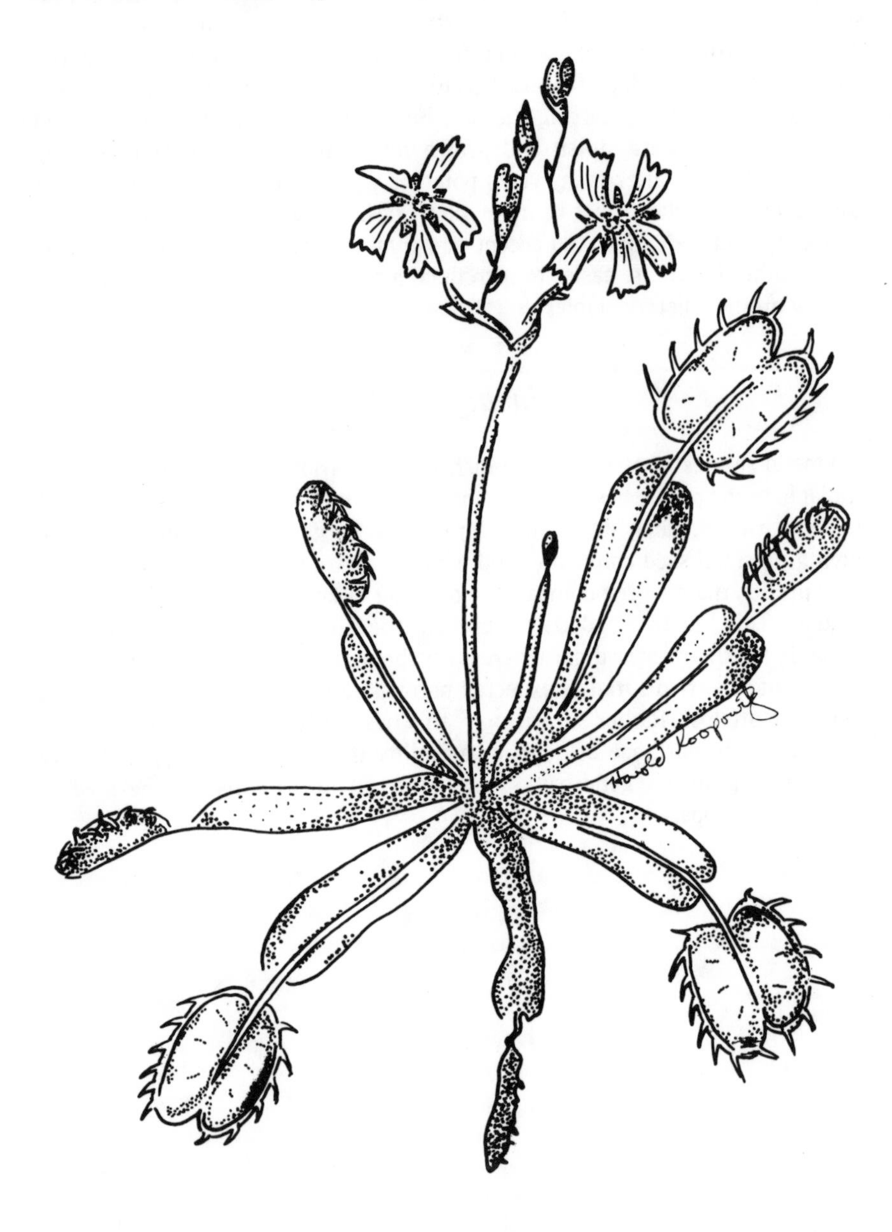

Case History:
The Venus Flytrap

All life ultimately is grass is just one way of stating that all animals eat plants or other animals that previously have fed on plants. Very few plants have turned the tables by trapping and eating animal predators. Probably the best known among the handful of carnivorous plants are the Venus Flytraps (*Dionaea muscipula*). These are widely available in nurseries and sometimes even in department stores, though they are facing disaster in the wild.

Venus Flytrap plants grow as a rosette of green leaves with each leaf divided into two portions: a narrow, flat stalk and a rounded, flattened terminal trap. Each trap, hinged along the center, is equipped with rows of teeth at the edges. Three small hairs stick out of the flat surface of each side of the trap. An unwary insect landing on the leaf and touching either two different hairs or the same hair twice will stimulate the hinge to snap shut and then be trapped inside as the unfortunate prey. The toothed edges of the leaf make it impossible for the insect to escape, and so eventually it dies. The trap secretes enzymes that digest the insect and then the leaf directly absorbs these nutrients.

The reasons for this curious evolution are obscure. The most common theory is that the plants live in nitrogen-poor bogs and so have had to adapt this unusual feeding technique to survive in these impoverished conditions. Animal protein is rich in nitrogen. Thus the Venus Flytrap may circumvent its nutritional problems by eating insects.

The Venus Flytrap already is officially considered to be vulnerable and seems to be rapidly heading toward its demise. The plants grow wild in the boggy areas of North and South Carolina where two different sources of danger threaten their existence: overcollection and encroaching forests. A ready market for living Venus Flytraps encourages their collection despite the fact that the plants have been protected by state laws for some time. Also, unscrupulous plant collectors may have falsified information about the abundance of wild plants in an effort to restrict the legislation. The only reason to harvest these wild plants is to earn greater profits since the cultural conditions needed to germinate the flytraps are well-known. Flytraps will live, grow, and even set seed quite easily—provided they are planted in living sphagnum moss, watered with spring or deionized water, and given adequate sunlight. Large scale commercial growing under these conditions is terribly expensive. Collecting wild plants can bring greater profits, and so the collecting continues.

Even if the collectors could be stopped, the wild population of flytraps stands little chance of surviving. The Carolinas have an aggressive reforestation program with bogs being drained and pine trees being planted. Few of the Venus Flytraps can survive the new, drier soil. Even these few plants are doomed.

Venus Flytraps will not flower and produce seed unless they are exposed to direct sunlight for several hours each day. As the pine seedlings grow they will shade the surviving flytraps with their branches and also smother the plants with their blanket of dense pine needles. In the shady canopy of the forest, the flytraps will not get enough sun to flower and subsequently produce seed. There may be no future generations of Venus Flytraps—at least not in the wild.

A relative of the Venus Flytrap, called *Aldrovanda,* also is in trouble. This European plant had leaf traps very similar to those of the Venus Flytrap, except that they were less than half the size. Because most of the marshy or boggy areas of southern Europe have long since dried, *Aldrovanda* probably is already extinct in its natural habitat. A few plants were fortunately transported to the Cape of Good Hope in South Africa and have become established there. Whether or not this unique species ultimately will be safe is still unknown.

CHAPTER 4

Industrial Plants:
Products, Fuels, and Fibers

Because plants are potent biochemists, man is able to obtain from them a wondrous assortment of industrial chemicals. Plants use these chemicals for their own specific purposes. These chemicals can be very valuable for man if we are able to understand, isolate, and manipulate these substances. Plants produce these chemicals as they absorb energy from the sun and convert that energy into a variety of products. Some products, such as lignin, are structurally sound and are used by man in various forms of wood. Many others, such as oils, rubbers, and resins, have other uses. While the list of natural products obtained from plants is extensive, the list of end products derived from these plant substances appears to be infinite.

Man already has discovered many uses for these plant substances. But in today's world where fossil fuels are dwindling and economic shortages are commonplace, many more significant uses are being considered for natural plant products.

At the present time, most of our energy needs are filled by fossil fuels—coal or oil. These fuels also provide man with an effective but costly chemical manufacturing base. Fossil oil is not a renewable resource. Eventually it will be depleted, possibly as early as the year 2000. As it grows more scarce, the price of fossil fuel will climb. In addition, politics often enter into the picture to cause sharp price increases, as occurred in the 1970s when the price of OPEC oil skyrocketed.

Living plants are a logical alternative to fossil fuels since coal and oil were derived from plants in the first place. The chemicals inside plants, called hydrocarbons, can be processed to form biocrude, which may be the answer to the high cost and dwindling supply of fossil fuels. Many plant species produce these hydrocarbons, but relatively few species produce the compounds in abundant quantities. Of those few high yield plants, only a relative handful of species contains products that can be easily extracted. As more and more plant species become extinct, we can expect to lose some of our opportunities to obtain new sources of biocrude.

Rubber Trees

Rubber is an industrial product that we use every day. While much of our rubber is a synthetic substance made from fossil fuels, natural rubber is a plant product. The milky latex that flows in the stems of certain plants has been understood for many years. Indians in the New World collected the sap from *Castilla* species, such as *C. elastica,* and produced bouncy rubber balls used in various games. *Hevea brasiliensis,* a species from Brazil, is the primary rubber tree used today. Even though this plant is a native of South America, most rubber plantations are located in Asia and Africa. The move away from South America occurred a little over one hundred years ago when a fungus, "South American Leaf Blight" (SALB), devastated the Brazilian crops. In an attempt to salvage the species, *Hevea* seeds were collected in Brazil and sent to Kew Gardens in England. More than 2,000 seedlings germinated at Kew, which were distributed to gardens in Java, Malaysia and Ceylon. Just twenty-two seedlings made it to Malaysia. These few formed the genetic base for today's commercial rubber plantations. These plants are relatively good rubber producers but carry no resistance to SALB. So far, the fungus has not yet reached Asia or Africa. If it does, we can expect the devastation of those rubber plantations. Meanwhile, the yield from these *Hevea* plants has been increased tenfold.

During World War II, chemists devised ways to synthesize rubber using fossil fuels. The quality of the synthetic rubber is not quite as good as natural rubber; thus most uses require mixtures of the two types. One reason that radial tires last longer than other kinds is that they contain a higher proportion of natural rubber. Tires made from just synthetic rubber are less expensive and wear down comparatively quickly. The natural product is consequently needed for aircraft tires and for heavy industrial use. American car tires average about 12.5 percent natural rubber, bus tires average about 55 percent, and aircraft tires contain 90 percent natural rubber. Synthetic compounds can be less costly, but they seldom compare in quality with the real substance.

Since we cannot rely on artificial rubber completely, an outbreak of SALB would be disastrous. To guard against this, SALB-resistant strains must be developed. Where will the resistant genes come from? In the jungles of Brazil, plant scientists have found a few resistant strains which have been infused into the Asian commercial stock. To date, complete resistance to the fungus has not yet been achieved.

Within the remnants of the tropical jungle, rubber trees may exist with genes for total SALB resistance and even higher quality rubber. These genes would be a bonanza.

Rubber from the Desert

Some plant families from the desert may prove to be even better sources of rubber, or at least attractive alternatives to the *Hevea* plants. Plants that produce latex and other hydrocarbons are found within a fairly large number of families ranging from annuals to perennials, shrubs to trees. A number of plant scientists are studying a variety of these species that may be valuable rubber-producers. One increasingly well-known species is *Guayule argentatum,* a member of the daisy family. Experiments are already being conducted with *Guayule* to test the quality and quantity of its rubber. A pilot rubber plantation has been developed in Mexico.

A lesser-known species, *Pedilanthus macrocarpus,* also called Candelilla, is being studied by Dr. Eloy Rodriguez, a plant chemist at the University of California at Irvine, and others. This species, which grows in Baja California, parts of northern Mexico, and the American Southwest, oozes a latex similar to that of *Hevea* and *Guayule,* in addition to a handful of other potentially useful hydrocarbons such as waxes and resins. Using species such as *Guayule* and Candelilla would be advantageous because they thrive in arid and semi-arid areas. Plantations could be developed in regions not fit for agricultural crops. Also, these plants would offer critical protection if SALB were to infest the *Hevea* plantations in Asia.

Living Oil Wells

Among other desert species being studied are some that produce substances similar to crude oil. Hydrocarbon fuels are made out of chains of carbon and hydrogen atoms. The very long chains with big molecules form natural rubbers, while the small chains with smaller molecules form oils. Nearly all plants that contain latex are actually miniature factories that pump out hydrocarbons of one sort or another. The milky white latex is a combination of water and hydrocarbons. When the water is removed, the remaining substance is biocrude.

A pioneer in the study of hydrocarbon-producing plants is Dr. Melvin Calvin, a Nobel Laureate on the University of California at Berkeley faculty. Calvin has done much to popularize the idea that we may be able to harvest hydrocarbons from plants and convert these products into valuable fuels. Calvin has done a great deal of work with *Euphorbia lathyris,* which is sometimes called the gopher plant because it is reputed to repel gophers. This plant yields ten to twenty barrels of oil per acre per year and it appears the yield can be increased through careful breeding. If a tenfold increase could be achieved, similar to that achieved for *Hevea,* this would mean up to two hundred barrels of oil per acre per year. The production costs for oil from the gopher plant are still rather high, but Calvin believes the cost could be decreased with further work. Since the future holds higher, not lower, costs of foreign oil, this plant may prove to be extremely important in the years to come.

One reason for switching away from fossil fuels is that we really do not know the extent of the problems we are generating by burning off coal and oil. Both fossil fuels are byproducts of photosynthesis, in which carbon dioxide and water are assembled into hydrocarbons. When the hydrocarbons are split back down into carbon dioxide and water, the stored energy from sunlight is released. Over hundreds of millions of years, plants have been quietly packing away this energy and using up the carbon dioxide.

The amount of carbon dioxide in the air in 1860 was 290 parts per million, a relatively low figure. Our burning of coal and oil has released carbon dioxide back into the air. Today, current levels of carbon dioxide are 330 parts per million. Scientists believe that as more coal and oil are burned the amounts of atmospheric carbon dioxide will climb even higher. A change from 330 to 400 or 500 parts per million may not seem like much, but this increase will have a profound effect. Sunlight coming through the atmosphere is absorbed by the earth and then radiated back into space as infra-red radiation. Carbon dioxide in the air absorbs infra-red radiation and re-radiates it back to the ground. As atmospheric carbon dioxide levels increase, we can expect a warming trend in the earth's climate. Such a trend (called the "greenhouse effect") would change patterns of agriculture, melt the polar ice caps, and flood coastal cities. By using fuel from living plants, rather than using fossil fuel, carbon dioxide would be recycled rather than added to our atmosphere.

Using Sugars

While some scientists are exploring the use of hydrocarbons, others have been testing the use of carbohydrates—such as sugars—as a base for making automobile fuel. The Brazilian government has begun harvesting sugarcane for this purpose. Using the fermentation process, the cane is converted to ethyl alcohol which is then mixed with conventional gasoline or water—either 20 percent alcohol to 80 percent gasoline (gasahol) or 95 percent alcohol to 5 percent water (ethanol). When water is used, a minor adjustment must be made to the automobile so that it can use the alcohol-water fuel. Besides the advantage of not requiring petroleum from fossil fuels, cars that burn ethanol are relatively non-polluting. In the United States, grain has been converted to alcohol to produce fuel, but there may be other species more suited to producing fuel oils. The search for plants that are rich in directly extractible fuels has moved into high gear. We may find that it is more economical to obtain crude directly from the plants without having to use fermentation processes. With this in mind, hundreds of species have been surveyed. Calvin has even located a tropical species (*Copaifera langsdorfii*) that he believes contains pure diesel fuel.

Additional Desert Helpers

Sometimes species that at first seem totally useless can turn out to possess important and novel characteristics. *Calotropis procera* is one such plant. A member of the milkweed family, *C. procera* is widespread in drier parts of Central and South America. People have tried unsuccessfully to use this plant commercially as a source of fiber, drugs, cattle fodder, antibiotics and even poison. Because this species produces latex, it was recently investigated for its biocrude potential. First reports indicate a productivity comparable to the gopher plant; the yield from the seed pods is even greater than that of the *Euphorbia* species. Refined breeding could make this plant an important addition to the energy arsenal being developed.

A number of desert species also produce resins, some of which may be converted to oil. The secreted resins usually form a sticky outer coating over the stems, leaves, and sometimes even the floral parts of the plant. One example is the Cresote bush, which yields extracts used as preservatives in margarine. Other resin-bearing plants, such as the *Chrysothamnus* species, produce hydrocarbons from resins that are comparable to those of the latex-bearing plants.

Poisons

Other plants provide us with poisons to use against pests. Our experience with DDT and other synthetic insecticides has taught us that the effects of these compounds can linger in the environment and be harmful to both humans and animals. Many of our safer insecticides today are based upon Pyrethrin, a compound produced by a species of *Chrysanthemum*. Being a plant product, Pyrethrin is biodegradable and is more specific for insects, not mammals. Since insects are one of the main enemies of plants, it shouldn't be surprising that plants have evolved chemical methods to battle back against insects. The very effective insect combatants that many plants contain have been exploited by humans only very recently. Rotenone, one interesting plant-derived poison, works on cold-blooded creatures but appears to be relatively safe for warm-blooded animals.

Chemicals from some plants are quite complex. When a leaf from a tomato plant is wounded, the leaf produces a protein that interferes with the insect's digestion. A few members of the tomato family produce chemicals that "turn off" the feeding behavior in caterpillars. Other plants secrete chemicals that stunt the growth of a predator. Plants in the carrot family produce a chemical that combines with the natural DNA of the insect to cause chromosomal abnormalities. Phytoalexins are fatty compounds within plants that have anti-bacterial and anti-fungal properties, and sometimes are thought of as the plant equivalent to interferon. Since some of the plants that contain large quantities of potent poisons are desert-dwellers, natural biodegradable insecticides could be made from these species that would effectively keep away insects from the new energy plants such as Guayule. Within the Compositae or daisy family, there is a group of compounds called polyacetylenes. Some of these compounds are more potent than DDT and could be used to combat mosquito larvae. Since malaria is on the rise again, these compounds may have considerable value.

One scourge of mankind is the freshwater snail that carries the intermediate stages for a wide variety of internal parasites. Bilharzia, a small blood fluke, has devastated much of the world where extensive irrigation systems are built. Unfortunately it is difficult to find effective molluscicides that will not kill off the entire stream ecosystem, including birds and fish. Scientists are now screening a variety of plants, hoping to find a safe molluscicide. More than two hundred native plants of Puerto Rico recently were screened for molluscicidal activity. Of these, *Solanum nodiflorum* was found to have compounds effective against snail species that carry both blood and liver flukes. The toxic effects on the environment still need to be investigated; so far scientists have found that this molluscicide does not kill other snail species. With luck, the compounds within *S. nodiflorum* may prove to be specific for the parasite-carrying water snails.

Most garden snails are *Helix pomatia,* a European import that has been spread from Europe in plant materials and carried around the world. We cannot help but wonder whether some European plant species possess an exceptionally potent molluscicide that could help rid worldwide gardens of their pesky snails. Is such a wondrous species alive or has it already succumbed?

Forest Products

The forests of the world provide us with fuel, wood to construct buildings and furniture, and fibers to produce paper and synthetic products such as rayon. Declines in the forests have already caused some repercussions in each of these areas. We undoubtedly will see more problems in the years ahead.

While timber is used for fuel in the poorer countries, people in the more affluent countries primarily use timber for construction and furniture. Wooden furniture prices continue to climb as the number of trees declines. In many cases, solid wood has been replaced with pressboard or veneers, both of which are less durable and less attractive. Consumers now do not have as many types of woods to choose from as our ancestors did.

We also have seen price increases related to paper and similar products. Concern over the diminishing forests has prompted serious reforesting efforts in many parts of the world, but these substitute forests are just not the same. Mixed forests that are rich in diversity have been replaced with monoculture forests made up of all pine or all eucalyptus. A monoculture forest certainly is better than no forest at all, but we shouldn't be misled by claims from the forest industry that these substitutes are equivalent. This concern is particularly important in the tropics where governments usually fail to differentiate between natural forests and monoculture plantations.

Dr. William Meijer, a conservationist, related a disturbing experience from the Conference on Improved Utilization of Tropical Forests held in Madison, Wisconsin in 1977. A group of conservationists pleaded for rational use of the forests. When it came time to publish the conference proceedings, their papers had been omitted. According to Meijer, the majority of the papers presented at the conference dealt with topics such as the virtues of machines that convert even small branches and limbs into chipboard or paper pulp, and the conversion of natural forests into monoculture pine plantations. The economics of forestry obviously overshadow the concern for the world's ecology.

Monoculture forests are not comparable to natural forests. They have no diversity because modern tissue culture methods have enabled us to clone individual trees. An entire forest can be replaced with genetically identical "supertrees." This undoubtedly makes good economic sense in the forestry world, but to have a forest of identical trees means there is identical resistance to disease throughout the plantation. Pathogens could easily wipe out an entire population. The same arguments against genetic conformity in crop plants hold true for forests.

Trees produce chemicals just as plants do. One frequently used chemical which is used in the production of paper and in the sizing of textiles is gum. Gums have a variety of other uses, ranging from food stabilizers to sizing plaster walls. Much of the world's gum supply is gum arabic which is tediously collected from species such as the *Acacia senegal* trees in the Sudan. The thick sap in these trees produces a gum when the trees are injured. The gummy sap works just like a scab to help cover the wound. Industrial demands for gums are increasing, but expensive hand labor is required to harvest these chemicals from the trees. Cheaper sources of gums definitely are needed.

In the search for better gum sources, guar gum was discovered toward the end of the 1940s in the seeds of *Cyamopsis tetragonolobus*. These plants are now raised for their gum content. About 50,000 tons of guar seed were imported into the United States in 1979. The demand is increasing . This is another example of an inconspicuous species that probably would have been overlooked if people in earlier decades had chosen which plants to save for the future. No one would have suspected prior to 1940 that this ancient legume could play a significant industrial role.

Fibers for Clothing

When synthetic fibers such as nylon were first introduced, the world markets for natural fibers, such as cotton and wool, plummeted. However, as with synthetic rubber, man discovered that the best clothes are made not out of natural or synthetic fibers, but out of blends of both types. Natural fibers once again are popular, which is fortunate because oil prices have boosted the costs of manufacturing synthetic fibers. Plants make many kinds of fibers that can be converted into cloth. While cotton is the most widely grown, many others are available, too. Linen, for example, is produced by the flax plant, *Linum*, which also provides linseed oil. *Cannabis* was widely grown as a source of fiber for making hemp long before its more popular recreational use became widespread. And sisal, from *Agave* species, is used to make ropes. Other species have the potential for producing tough, yet flexible, fibers.

Perhaps the most difficult part about relying on plants for these kinds of products is that we can't predict which plants possess valuable products or when a product will be needed as a substitute for something in scarce supply. Whales were first being exterminated mainly for their oil—an excellent lubricant for heavy machinery. Now we have learned that *Simmondsia chinensis,* a desert plant, has seeds that contain oil that is almost an exact replica of sperm whale oil. *Simmondsia* plants are now being grown on plantations to provide jojoba oil which can be used in a variety of products. This discovery of jojoba oil may have come too late to save the whales from eventual extinction, but we do know that it will save *Simmondsia chinensis* from being allowed to fade away.

Fitzroya cupressoides

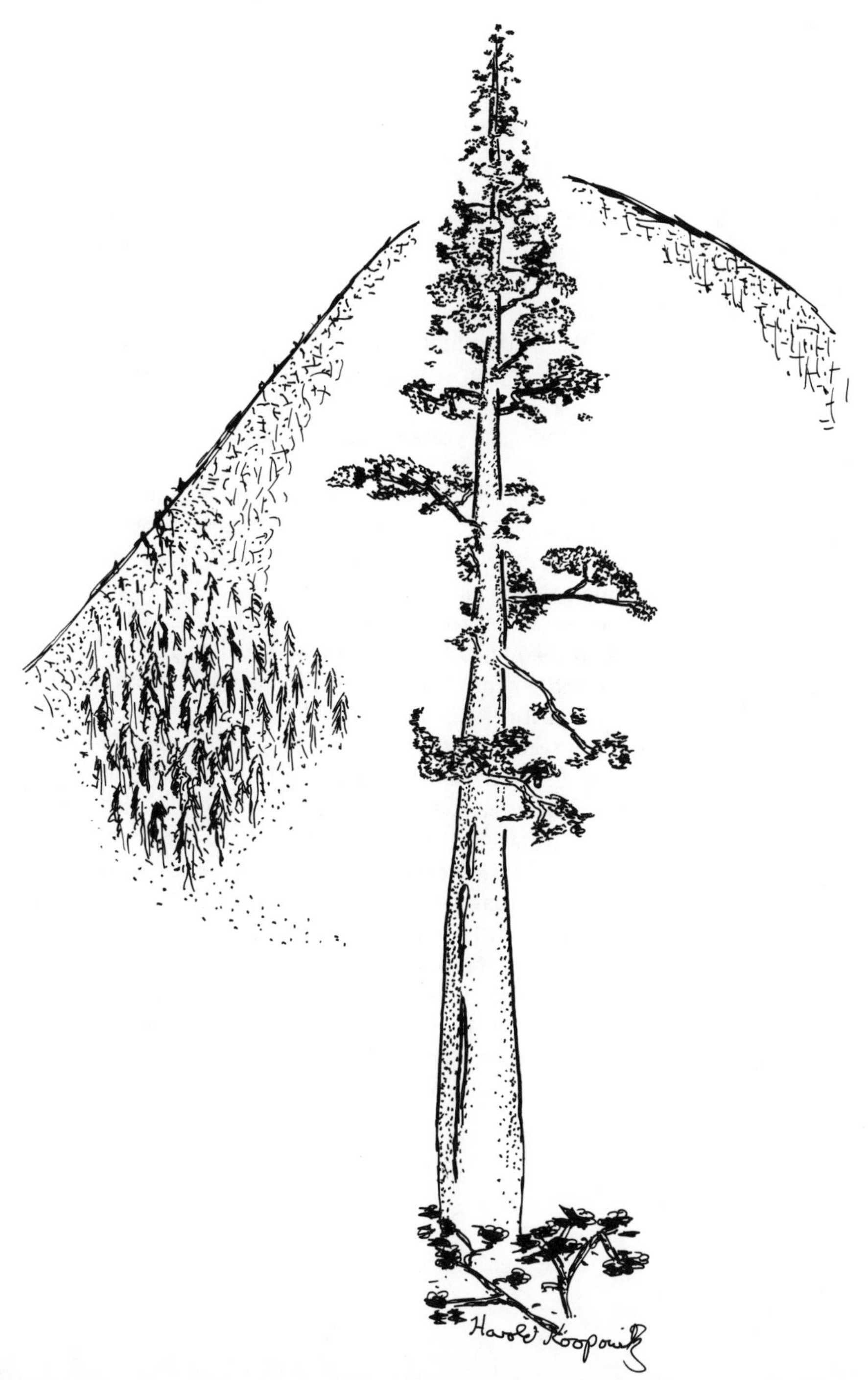

Case History: The Giant Alerce Trees

These giant trees reach nearly two hundred feet into the crisp, Chilean sky. Their massive trunks can spread well over twelve feet in diameter and are girdled in a fibrous, reddish bark. These are Alerce trees—considered one of the best woods for musical instruments but frequently used instead for roof shingles. These giants, properly called *Fitzroya cupressoides,* resemble the towering redwoods that grow along the Pacific Coast of North America. Both are conebearing trees, and both thrive in the cool, moist air near oceans.

Alerce wood is durable, resisting fungi and insects. A fallen trunk can last for centuries before succumbing to natural decay. The trees are very slow-growing and cannot be quickly replaced after they have been cut. Alerce's ability to persist for two to three thousand years makes their destruction particularly sad.

Alerce trees have been chopped down by man since 1599 and lumbering continues today, despite "protection" in both Chile and Argentina, where these trees are usually found. Extinction warnings were first raised back in 1896 but no one paid any attention at that time. In Argentina, Alerce trees exist today merely as remnants, somewhat protected in national parks.

Most of the damage to the Alerce forests was completed by 1850. The end of the colonization and settlement of southern Chile marked the most rapid period of deforestation in Latin America. These slow-growing giants do not regenerate easily and few seedlings germinate in the logged areas. The only forests that still exist in Chile are small patches of trees growing in boggy lowlands, or at altitudes between 2,100 and 3,000 feet. Several legislative attempts have been made to save these trees but none have succeeded. Laws were passed in 1969 to protect the Alerce trees. In 1976, the species was declared a Chilean national monument, making the trees theoretically exempt from exploitation. The species now is listed on Appendix II of CITES, which specifies that trade in Alerce lumber is forbidden.

Despite all of these well-meaning laws and regulations, logging has continued in Chile to a large extent. The bottom line is that Alerce trees are of prime importance to the economy of southern Chile. In the end, this is what seems to matter most.

CHAPTER 5

Ornamentals and Aesthetics

There is obviously much more to life than the basic needs of food, clothing, and shelter. One of our universal traits is to surround ourselves with greenery. Psychologists have even suggested that people find green to be a soothing color. Even those without gardens enjoy accumulating pots filled with plants and vases filled with cut flowers. A modern emphasis is placed on landscaping—whether the greenery adorns the edges of a freeway, a big municipal park, or simply the entrance to a suburban condominium. We've already seen how the extinction crisis may affect specific areas of our lives, such as the food we eat and the medicine we take for our diseases. But how will the crisis affect us on an everyday basis?

A big concern is that many of our popular potted plants are natives of the tropical forests. The warm, even temperatures and the dim light on the damp forest floor are similar to the conditions found in our homes. *Pothos, Philodendron,* and *Syngonium*—all of these and their relatives make ideal, hard-to-kill plants. In their native habitats, these plants search the forest floor, looking for a tree to climb into the sunlight. In our homes, they twist and turn in their pots as they try to find the tree that will carry them into the sun. Such forest-dwellers can survive with limited light from lamps and windows.

Plants undergo various cycles of acceptability. During Victorian times both palms and aspidistras enjoyed great popularity. Yet during the middle of the twentieth century both fell from favor. In recent times the palm has once again become a favorite, but aspidistras are rarely seen in homes. Tropical forests have provided many exciting and novel plants through the great centuries of exploration, but we have scarcely scratched the surface. Many more species exist in the jungles that have never been seen. The steady decline of the tropical forests will directly affect both the numbers and the kinds of house plants sold in our nurseries, plant stores, and supermarkets.

The Big Flower-Breeding Business

Since flowers are an important part of our lives, flower breeding is big business. Not content to enjoy flowers in their natural form, flower breeders carefully mix together different species to enhance the bouquet. Flower hybridizers usually follow a general pattern when they breed variations. The breeder's initial attempts are to mix species to produce bright, clear colors and the biggest possible flowers. The results are six-foot gladiolus spikes with florets five inches in diameter, or perhaps dahlias the size of a dinner plate. These are truly exhibition flowers, but aside from their prize-winning traits they have very limited use. Exhibition flowers spoil easily in bad weather and are too large to be used as cut flowers. Ten-inch orchids may be wonderful in competition, but there's hardly a place for them as corsages. Most women would be dwarfed by wearing such massive flowers pinned to their chests.

For nearly all garden flowers, the next breeding stage is the creation of dwarf forms for use as bedding plants and to yield the smaller sprays that are perfect for cut flowers. Dwarfness generally is the result of breeding selections made within the hybrid stock. Daintiness, however, is a prized trait that is best produced by introducing wild species back into the hybrids. Wild species help create the most novel kinds of flowers.

Flower coloring is an interesting science in itself. After producing bright, clear colors, hybridizers usually turn to the creation of more subtle shades and colorings. Breeders frequently try to achieve naturally rare colors, such as black or the elusive blue shades. We look forward—although with some trepidation—to the day in the future when genetic engineers can take the blues from delphiniums and put them into roses.

Unusual Gladioli

At the University of California, Irvine Arboretum and Gene Bank we have a very large collection of gladiolus species—approximately half of the two hundred known species. Most of these are wild species and unfamiliar to most people. Garden gladiolus, the kind we find in our gardens and florist shops, have been bred from just six species and have led to thousands of man-made varieties. Because the variety appears to be endless, many people assume that all possible varieties have already been created. Not true! In the wild there are dwarf species scarcely six inches tall, many deliciously scented versions, species that grow and flower during any season, and many others with unique traits. At the arboretum we are exploring this vast potential and are creating garden hybrids quite unlike anything else ever produced.

One currently threatened species is *Gladiolus guenzii,* which grows as an evergreen clump on sand dunes in South Africa. The leaves produce a graceful fountain effect and the small flowers bloom over a long season. If large, colorful flowers were to be introduced into its genetic makeup, *G. guenzii* could make a good perennial landscaping plant for warm climates. We had grown this species in Irvine for years and were surprised to find it, and others familiar to us, on the endangered species list for South Africa. As more and more species are threatened, botanic gardens will become the unwitting guardians of endangered species.

Although common gladiolus plants flower during the summertime, most of the wild gladiolus species flower in the winter. These plants hail from Mediterranean-like climates where winters are mild. New, winter-flowering, garden-variety plants have recently been produced by incorporating the genes from the wild *Gladiolus tristis* into the big, garden hybrids. While these new varieties are useless for gardeners in harsh climates, they could be potentially important crop flowers in areas with mild winter climates such as Southern California. Another winter-flowering plant, *Homoglossum priori,* is a bright red and yellow cousin of gladiolus that can interbreed with gladiolus species. Because it flowers in late November and early December, it has the potential to produce a new Christmas flower in the Northern Hemisphere.

The scents of wild gladiolus are varied—ranging from scents that resemble clove-carnation to the finest French perfumes. A few unusual species produce a strong, sweet, freesia-like scent that can be enjoyed only by those people who are genetically disposed to detect this fragrance. Those who have not inherited the "right nose" are unable to smell the flowers at all. A casual poll taken at the arboretum revealed that about 75 percent of the population has the correct gene to enjoy the scent.

Truly awesome potential resides in some varieties of *Gladiolus liliaceous.* Not only are these flowers spectacular but they are reported to change color from morning to night. The flowers are creamy-yellow in the morning. By nightfall the blossoms have turned to a pinky mauve. If the genes controlling this trait were selected, improved, and then introduced into larger flowers, we could create flowers that change color in the vase!

One of our favorite species at the arboretum is *Gladiolus citrinus*, a very rare plant that is now almost impossible to find in the wild. At one point in time, when *G. citrinus* could not be located in the wild, plants were finally found growing in a French hobbyist's collection. Seed was sent back to wild flower enthusiasts in the plant's native South Africa and, in turn, we obtained the arboretum seed from them. *G. citrinus* has bright yellow, crocus-shaped flowers on relatively short stems. We believe this species can be used to breed a dwarf edging gladiolus. Some scientists think *G. citrinus* is very primitive and close to the ancestral species that evolved into the large gladiolus group. If so, the preservation of this species carries a special responsibility.

The list of possibilities continues. Some species have interesting spottings and markings, while others have strange shapes and flowering habits. We even wonder if the corms could be used to make a new food crop, since one species, *Gladiolus edulis,* used to be eaten.

Fashionable Flowers

Fashion exists in flowers and plants just as it does with clothing, hairstyles, furniture, and everything else. Flowers frequently appear and disappear because of unidentified whims. At other times flower fashions are dictated by specific social mores. The Victorian era, when the British Empire was at its peak, was not a good time for yellow flowers. Yellow was the color associated with cowardice. It simply wasn't proper to have beds of yellow pansies, marigolds or tulips. Anyone with any sense would not lay a yellow swath in any British park. Only the British clergy, long noted for their eccentricity, seriously considered yellow flowers. They started breeding golden daffodils and eventually turned them into respectable garden flowers, the heralds of spring.

During the twentieth century, when it became chic to be daring and modern, colors needed to be bright, clear and strong. No one would tolerate a daisy, for example, in shades of dull magenta or a red rose with a purplish tinge. Today, perhaps because the future is no longer so bright and utopian, people "ooh" and "ah" over magenta Senecio daisies, bouquets of lavender alstroemaria and purplish roses. What will be the fashions of tomorrow? More importantly, how will we make those flowers if we save only the fashionable ones of today?

Tulips

Wild species have made tremendous contributions to modern garden flowers. The ever-popular tulip is a good example. Tulips have a lengthy and honorable history as garden flowers. Long before the Dutch got hold of the tulip-growing monopoly, Persians and Turks cultivated these brilliant bulbs and wrote poems about their beauty. When the Dutch started to grow tulips, they discovered that some varieties commanded higher prices than others. These premium bulbs became commercially available and therefore fashionable. Actually, those were rather dull flowers of white, yellow, or cream with maroon or brown markings. A virus disease inhibited the vivid colors from appearing. But since viruses had not yet been discovered, the notion that the flowers were sick did not surface.

Many different tulip fashions have been in existence over the last two hundred years, but it was not until 1936 that a striking advance was introduced. When the species *Tulipa fosteriana* was hybridized with Darwin Tulips, the result was gigantic, tall flowers. These Darwin hybrids, as they are called, have to a large extent supplanted most of the other kinds. The Darwin hybrids illustrated the powerful effect of crossing the correct species with a hybrid. Two other species, *Tulipa kaufmanniana* and *T. greigii,* exert totally different effects. They create hybrids with short stems—often barely six inches tall—and large flowers. This makes them an ideal edging plant for borders. As a bonus, these hybrids have leaves with maroon stripes and spots. Another species, *Tulipa retroflexa,* a rather scruffy-looking species, produces the very graceful "lily-flowered tulip" when crossed with the larger hybrids. The process of tulip hybridizing has not yet ended. Species are needed to produce better heat-tolerant tulips that will not require a winter chill to make long flower stems. Other species hold the genes to make truly perennial tulips that, once planted, will increase and clump, with the clusters coming back year after year.

Daffodils

Daffodils are another group of garden plants that benefits from wild species. Large garden varieties of daffodils have recently been crossed with many of the small wild species to produce delightfully graceful blossoms. *Narcissus triandrus* and *N. cyclamineus* have been used in breeding for many years. The first species makes small clusters of blooms with very silky petals. *N. cyclamineus* usually gives genes for flowers with backswept petals, a long-waisted trumpet, and early flowering capabilities. These types of daffodils are constantly popular. The demand for miniature daffodils far exceeds the supply. Unfortunately the ability to make new kinds of miniature daffodils is hampered by the disappearance of many of the tiny, wild species. *Narcissus calcicola,* a tiny yellow jonquil from Spain and Portugal, is considered endangered and *Narcisssus watieri,* possibly the most powerful tool for making miniature white daffodils, is already unobtainable. Occasionally the *N. watieri* from North Africa's Atlas Mountains is advertised by unscrupulous bulb merchants who substitute another variety. The status of *N. watieri* is unknown. The only hope is that a few plants may still exist on some of the rocky ledges or hillsides where the species once thrived. If so, perhaps it can be introduced once again. More likely, the species already has been destroyed—either over-collected or eaten by goats.

The Latest Improvements

New garden flowers can still be made from the species remaining in the wild. In the last few years we have seen the transformation of impatiens with species discovered in New Guinea. Plant breeders can now make impatiens with larger flowers and variegated leaves. A friend who hybridizes begonias has added the genes of a species with variegated leaves to the large, bedding tuberous begonias. The result is plants that resemble Rex Begonias but which have large, double waxy flowers.

One of the most popular introductions in the potted plant world is *Streptocarpus,* the Cape Primrose. There, success has spread in a step-like fashion as each species placed into the hybrid gene pool has added its own characteristics. In one step, *S. dunni* added red and pink tones. In another, *S. johannis* created the popular nymph series. Other species exist back in the wilds of Africa that could miniaturize the plants and even transfer a measure of frost tolerance so that the plants could be grown outdoors in colder climates. The secret to using wild species is knowing how and when to use them. How many building blocks will be lost in the next twenty years?

New Directions in Landscaping

As we drive through the coastal cities of Southern California, we can date the age of the community by the kinds of landscaping planted around the homes and along the streets. We can see distinct sequences: older communities are framed with pastel hydrangeas, the middle-age areas display the blaring reds of bouganvilleas and the youngest communities are draped in the softer pinks of *Raphiolepis.*

While much of the older landscaping relied heavily on water to survive the dry, hot summers of Southern California, the trend now is to find drought-resistant plants that can make it unaided through the hot season. Where will we find these new stalwart shrubs? People now are looking at native California plants as their new landscaping materials. If the move to drought-resistant plants occurs soon enough, we may see baccharis, yuccas, and erioganum clothing our freeway edges instead of daisies and other water-demanding plants that are currently used. The search is none too soon. California has one of the largest concentrations of endangered species in the United States. More than two hundred of the state's endemic species are threatened in some fashion or another. This state of affairs is partly due to the exceptionally large numbers of naturally rare endemic species. The Golden State's population boom is another culprit.

Antique Varieties

Just as general agriculture has experienced a decline in species variety, the same trend is apparent in ornamental horticulture. Thirty years ago seed, and plant catalogues offered tremendous variety. Today, handling small quantities of a plant is so expensive that few companies want inventories with hundreds of different cultivars and varieties. In 1929 the renowned bulb firm of J. Scheepers offered forty-one varieties of crocus, of which sixteen were wild species. In 1977, the offerings were down to 11 varieties and none were wild species. The pattern is similar for many groups of plants. Larger companies generally have fewer varieties. Only the smaller businesses, where love of the plants rather than money is the motivating factor, will carry some of the more obscure varieties.

Is it important to save antique varieties of garden plants? Yes. Changes in fashion can make antique varieties popular once more. Take orchid plants as an example. During the latter part of the last century, hybrids were made between selected, wild species of *Paphiopedilums,* the slipper orchids. A hybrid between two species is called a primary hybrid. Hybrids between primaries are complex hybrids. Most modern slipper orchids are the result of several generations of complex hybrids. Although the primary hybrids of the last century had a special charm of their own, they do not compete with the flashy complex hybrids. The fashions during the 1940s and 1950s—which mandated large, rounded flowers with strong colors—favored the complex hybrids. In contrast, primaries tend to be elongated flowers—often with fanciful shapes—and the colors are often muted with unfashionable maroons and somber browns. During the last fifteen years, the infatuation with large complex hybrids has begun to wane and has been replaced with new interest in species and primary hybrids. Unfortunately few primaries still exist from the previous century because they had become so unpopular. Many of the species luckily were still available. Hybridizers began remaking the old primaries. Some species, like the *Paphiopedilum sanderianum,* are now so rare that their existence has been questioned. Obviously their primaries cannot be remade. The *P. sanderianum* had petals at least eighteen inches long and could have contributed to fantastic flowers. It is believed that *P. sanderianum* exists in a preserve near Mt. Kinabalu in Borneo, but it cannot be used to remake the earlier primaries. What will happen if primaries again disappear and hobbyists once again want to remake those flowers? Will we have any species left as raw material?

Another example of the need for old primary hybrids comes from daffodils. Tazetta daffodils are bunched narcissus flowers, very popular at the turn of the century. They made excellent cut flowers. Since they were not very hardy, they lost their popularity to the larger daffodils that could be cultivated in colder climates. As daffodil growing has spread to warmer climates, tazetta daffodils have become popular again. Many older varieties are being rescued from old gardens but nearly all are sterile and useless for breeding. For a while it appeared that significant tazetta breeding could not occur. Breeders have finally uncovered a very old variety called Gloriosus, a sweetly-scented flower, growing in the Scilly Islands off Cornwall, England. These flowers proved to be fertile. Now, by mating Gloriosus to large, modern daffodils, we may be able to create a totally new race of daffodils.

Flower breeding is an unusual blending of the arts and sciences. Indeed, this is one of the few areas where the two worlds meet. The latest scientific methods are frequently used to create and propagate new varieties. However the success of a new flower is ultimately based on aesthetics. The biggest question is, does the new flower have popular appeal? What the sciences may create, the arts will then assess. Neither world will be able to overcome the problems that will result if the present extinction rate continues to accelerate.

Begonia socotrana

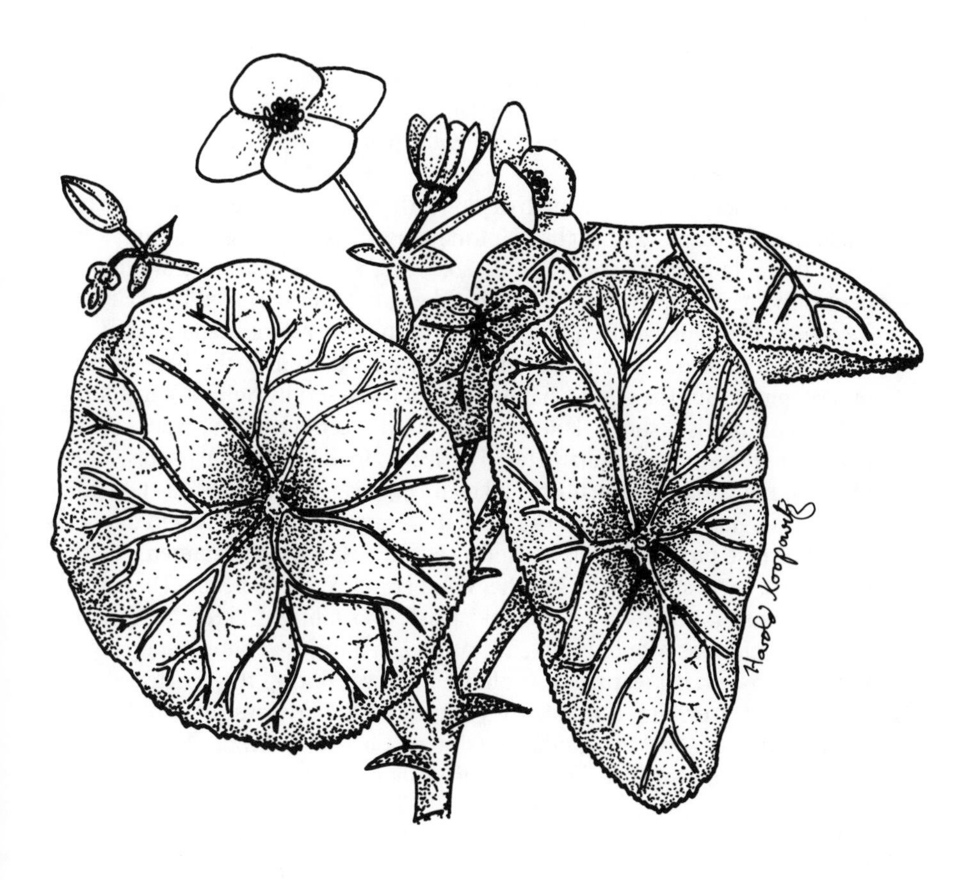

Case History: The Bulbous Begonia

The magical tales that used to enchant us when we were young were often filled with stories of buried treasure—quite often treasure buried on remote desert islands visited by swashbuckling pirates. This case history also concerns buried treasure but it is not the fine caskets of precious gems nor barrelsful of gold bullion in our dreams. The treasure we speak of is *Begonia socotrana* from the island of Socotra in the Indian Ocean off the Arabian coast. The 70-mile-long island has been home to nomadic peoples and their legions of goats for many, many years.

Begonia socotrana is a living treasure with genes so valuable that they have made many fortunes within the plant industry. This species grows from a true bulb, the only begonia species to do so. Other species also have swollen underground parts but those are actually tubers, not bulbs. The bulbs of *Begonia socotrana* hardly look like a king's ransom. In their native island habitat they were small clusters of spheres buried in the sandy soil at the base of rocky outcroppings. But the plants that grow from these scruffy bulbs are delightful.

B. socotrana is a short plant bearing round green leaves that vary between four and seven inches in diameter. The leaves look like little parasols, rather like those of garden nasturtiums. Like all begonias, this species has two kinds of flowers—male and female. The male flowers are an inch or two in diameter with four rose-pink flat petals around a central boss of golden stamens. Female flowers are fewer and smaller but they boast six narrow pink petals. The flowers, born in clusters, create a pretty pink splash over a bright green background.

This species was brought into cultivation just a little over one hundred years ago. Its most prized feature is its ability to flower during the shortest days of winter, thus providing a wonderful contrast to the dreariest days of the cold season. What cheer this would bring during Christmas. While the plant itself was popular, the two hybrids it produced soon surpassed the parent plant in popularity. Both hybrids have proven to be quite valuable over the years.

The first hybrid was created in 1883 when *B. socotrana* was crossed with a tuberous begonia hybrid that bloomed in the summer. The offspring inherited the winter flowering traits of *B. socotrana* and the physical characteristics of the other species. A number of different hybrids were introduced and these were lumped together as the Hiemalis (winter) group. A major advancement in these plants occurred in 1955 when Rieger introduced new "improved" Hiemalis begonias. These plants were patented and today they enjoy worldwide popularity as potted florist flowers.

A second group of hybrids was created in 1891 when the French crossed *B. socotrana* with an African tuberous begonia species called *B. dregei.* The resulting plants, called the "Christmas Begonias," had so many white or pink flowers that they frequently hid all of the plant's foliage. One variety, Gloire de Lorraine, has remained in cultivation and is popular today but, as a group, Christmas begonias never achieved the popularity of the Hiemalis begonias.

B. socotrana still contains the potential to breed new types of winter flowering begonias. What a shame if this species and its valuable genetic potential is lost. On the island of Socotra the population of begonias has steadily eroded. Many of the island's plants are now listed in the Red Data Book. Of the 216 endemic higher plant species on the island, 132 species are considered to be in danger. Of these, 85 species are on the verge of extinction. In most cases, poor husbandry is responsible. The natives have let grazing animals roam uncontrolled. This results in the depletion of the begonias. Conservation efforts are not in effect on the island and most, if not all, of these species will be lost. Unlike the related African floras, the plants of Socotra evolved without the presence of grazing animals. Many species are thus unique. When this island's unique species disappear, they will be lost forever.

II.

EXTINCTION:
WHY AND WHERE

CHAPTER 6

Biology and Extinction

Extinction is a one-way street—a well-traveled road followed by all groups of organisms. While mankind is often blamed for the extinction of other species, this usually is not the case. The disappearance of an entire species usually is a natural process. Many biologists believe that only a fraction of one percent of all the organisms that ever existed still are alive today. More than 99 percent of all the species that ever walked, crawled, flew, or swam on earth are gone. For most groups of organisms, extinction is the eventual, inescapable conclusion to life.

If extinction is so natural, why all the fuss? Simply because extinctions are suddenly occurring almost faster than we can count them. Unnatural extinctions—the kind that occur when a species meets a premature demise—are much more common today than ever before. We may not know why the hulking dinosaurs were wiped out, but we surely understand what happened to many other species. We're certainly well-aware of mankind's role in the disappearance of species today.

Scientists have categorized two main forms of extinction: true and false. In true extinction, the species simply dies out because it is unable to adapt to natural or unnatural changes in the environment. In false extinction (also called pseudo extinction), the species manages to adapt to the new environment but, in doing so, evolves into a new species. This extinction process is gradual but the end result is the same: the disappearance of the old species. True extinctions are the most prevalent, making up at least 80 and perhaps as much as 95 percent of extinctions. In this chapter, we will explore some of the different aspects of the extinction process that help explain what is happening in the plant kingdom and what those events may mean for us.

When is a Species a Species?

To understand the biology of extinction, you must grasp a few basic facts about species and evolution. Understanding what we mean when we refer to a species would be helpful but, to be honest, biologists for decades have unsuccessfully searched for an adequate description. The definition we use—that a species consists of individuals that all possess the same body form—is the most popular definition among biologists, but even this is sometimes confusing. The rather simple-sounding definition can get quite complicated. Individuals may look alike but, when we measure their various characteristics, we sometimes realize that the individuals are not identical after all. Some animals and plants have differences that are difficult to spot but which definitely indicate a different species. Humans often miss these differences in other species simply because most of us are not trained to discern such minute differences. This unfamiliarity and lack of training is the same phenomenon that occurs when humans from one race have difficulty seeing variations in individuals from another.

Because identifying a species using only physical characteristics can be difficult, even scientists have trouble agreeing on who's who and what's what. Some scientists, weary of grappling with these matters, advocate throwing out the species concept altogether. This would be a mistake because, despite the confusion, separating organisms into different species is a valuable way to pigeonhole and categorize organisms. Zoologists sometimes use a different yardstick to measure organisms. They define species by grouping together individuals that can reproduce among themselves or with other, similar populations. These groups have developed genetic barriers to reproduction with other groups and, if offspring are produced, they are sterile. The most familiar example is the mule, the sterile offspring that results from the mating of a horse and an ass.

Plants cannot be grouped in this same manner. Barriers between groups of plants are not genetic; they are governed by pollinators—bees and other insects—that visit all flowers of the same species during a foraging run. Since genetic barriers do not exist within the plant, horticulturalists can create fertile plant hybrids using different species. In the end we are left with our old, slightly imperfect method of identifying species via physical characteristics.

Survival of the Fittest

The individual variations we see in members of the same species frequently are quite logical when viewed in the context of evolution. Charles Darwin observed in the nineteenth century that not only were different populations of organisms variable, but these organisms often produced more offspring than could survive within limited resources. Darwin hypothesized that those organisms that had favorable variations held an edge over their peers and thrived. The offspring of the survivors inherited those favorable variations and they, too, survived. This was Darwin's famous "survival of the fittest" evolutionary theory. As the variations become more and more defined, organisms eventually diverge from the parent stock. This evolutionary adaptation leads to false extinction in extreme cases.

In a very large population—whether it's humans, monkeys or marigolds—many forms of stress complicate life. Variations evolve to cope with these stresses; often, more than one evolutionary variation is used to cope with a single problem. As a simple example, look at a plant population usually pollinated by bees. For our purposes, imagine that the bees were wiped out by disease or insecticide. Without the bees the plant's reproduction would be seriously threatened. Flower color is one variation that could help some of the plants. A butterfly population that prefers red and orange blossoms may migrate into the area. These butterflies would pollinate only those plants with red or orange flowers and only these plants would survive. Before long this plant population would have only red or orange flowers. In another hypothetical situation, pollination might occur when a night-flying moth visits and pollinates flowers that can be seen in the moonlight. In this situation, plants with white flowers would be the evolutionary favorites because their flowers are most visible at night.

These sorts of evolutionary changes frequently occur in nature. The end result is the creation of many groups of individuals that are different from each other but usually not distinct enough to be considered a different species. These are called sub-species or incipient species, meaning that they are in the process of evolving into a new species. While some sub-species may wind up as a distinct, new species, others eventually can be reabsorbed into the original population and still others may die off. Because sub-species contain traits different from those found in other members of the species, saving these groups may be just as important as saving the species itself. If a sub-species becomes extinct through unnatural causes such as overpicking, a potential new species is eliminated even before it has a chance to exist. The endangered and threatened lists show that this is happening right now.

Genes versus Environment

The variation seen in organisms results from interaction between two different systems—genes and environment. Genes, those inherited parental traits, create a set of blueprints that specify how an organism will be built and how it will work. Environment encompasses both the space in which a plant or animal lives, and the external events that occur in the organism's life. The interaction of the environment with the genetic material produces a distinct individual. The environment's impact may be rather insignificant or it may be tremendous. To see how an environment can produce devastating results, look at human populations in underdeveloped countries. Roughly 60 percent of children in poverty-stricken nations suffer from malnutrition. Although these youngsters have been programmed through their genes to achieve certain mental and physical capabilities, the environmental effect of malnutrition frequently will block these goals. Similarly, man's physical stature has changed dramatically since medieval times. As measured from old suits of armor, the height of knights often measured less than five feet; a man five feet six inches was considered tall. Today's average man is taller and the mean height is increasing. While human genes contain this capacity for greater height, environmental factors such as healthier diets and improved medical care affect these genes.

Environmental influences are just as important to plants as they are to humans and other animals. We can see the environmental impact in an example from the Sierra Nevada mountains in California. There, at high altitudes, several spectacular species of delphinium grow. The seedling grows a rosette of leaves during the first year and then becomes dormant beneath a winter mantle of snow. A warm spring thaw permits the growth of a central bud that produces a single spike of violet-purple flowers, sometimes five to six feet tall. These vivid flowers line the sides of streams and riverlets in the late spring. If a hungry caterpillar happens upon the scene and chews off the top of the bud, the delphinium will not flower until autumn. At that time the plant will produce several, shorter spikes with smaller, less spectacular flowers instead of growing a single, tall spike. A naive plant collector would assume that he had come across two somewhat similar species instead of one species that was greatly affected by the environment—in this case a hungry caterpillar. Environmental differences will affect even the genetically identical cloned plants. The Peace Rose which grows in the dry shade on one side of a house will be similar, but not identical to the fat, well-fed Peace Rose that makes its home on the sunny side.

Vanishing Genes

All this talk about genetic variability may seem far removed from the main topic of extinction, but variability is the key to the concern over the escalating extinction rate. When species become extinct, their genetic variables also disappear. Vanishing genes make the extinction problem more worrisome than most people realize. A smaller number of genes means less variety in the plants that may evolve or survive. Variation has always played an extremely important role in agriculture. Crop experts for years have pushed for the formation of seed banks to retain "antique" varieties of vegetables. Disease could wipe out a particular, vulnerable variety—a plant that may have had disease-resistant ancestors but which no longer carries those genes. When the older, disease-resistant varieties are lost, farmers have nothing left to plant when their crops are destroyed by disease. Outbreaks of new and virulent plant diseases are no bogey man—they do occur in unpredictable and devastating forms.

Life's Blueprint

Most animal and plant species share the same genes that create the essential blueprint for the biochemical reactions that underlie life. We are interested in the tiny remainder of the genetic material which carries the codes for subtle variation; the genes that program the organism to be short or tall, pale or dark, or any of a myriad of other little details. Most of these traits have little importance for the organism unless the environment suddenly changes. If such a change should take place, these details may be vital to the plant's survival. For instance, a plant may contain genes that carry resistance to a particular disease. While these genes are unnecessary if that disease does not occur, they are tantamount to survival if a plague breaks out.

Variations occur within a species because not all genes are expressed by an individual. An individual may carry genes which aren't expressed until succeeding generations, or it may have duplicate genes for a given characteristic. Other individuals may carry several different genes influencing a single characteristic and only one is expressed. Some snapdragons, for example, carry genes for white and red flowers but express only white flowers.

The larger the population, the greater the number of genes the population might carry. Geneticists refer to all of the possible different genes that exist in a population as a gene pool. Big populations have large gene pools and small populations have restricted gene pools. Unfortunately, a small gene pool often limits a species' ability to adapt to an environmental change. A scientist looking for a particular gene—perhaps cold hardiness or heat tolerance—would have a greater chance of finding it in a large pool than a smaller pool. Large numbers of genetically different individuals have a far greater chance of preserving the species.

Mutations

Creating new genes through the use of mutation may sound like a useful alternative to preserving a large gene pool, but mutants are unpredictable.

Mutants are created by bombarding genes with high energy radiation or using special chemicals called mutagens. When this is done, most of the obtained mutations are deleterious. Predicting which genes will be changed is nearly impossible. Scientists may have to make thousands of mutations and then screen these to find the desired one. Even then, chances are slim that the correct gene has been changed and that the change is what was sought. This process has been successful in only a very few isolated cases. Even with a very detailed knowledge of the biochemistry within the cell being altered, scientists still do not know how to affect the exact change in the gene they need. If they did, the possibilities would truly be endless. Imagine the economic value of a sky-blue rose. Scientists already know enough about the pigmentation of flowers to realize that a single mutation in the genes that produce true red pigment would change the color to true blue. However, science is still generations away from being able to perform these kinds of miracles. Genetic engineering, despite all the hoopla you've heard, is still in its earliest infancy. Instead of expecting miracles from genetic engineering and artificial mutations, we should preserve and utilize genes already present in the wild. Gambling on the ability of future scientists to produce effective engineering is too great a risk.

What Genes should be Saved?

Although choosing which genes to save is a difficult task, there are a few guidelines to follow. Some considerations deal with differences in stature and shape of plants. While short plants are most popular today, tall plants might come into vogue in the future. We also need to conserve tolerance to colder and hotter temperatures to permit plants to be grown in the more extreme climate zones. Future plants must also contain higher protein levels. Looking at the need for more protein, we see that ten pounds of hay are needed to sustain each pound of cow, the source of much of our protein. If enough nourishment could be gained directly from hay, ten times as many people could be fed at much less expense. This would be a tremendous advance in the efforts to reduce world hunger.

Plants that can grow in salty water and under arid conditions are as important as novel kinds of food crops. Many of these new food sources will come from breeding familiar food plants, but some new crops may turn out to be what are now wild, but edible, species. Edible wild plants, even though they are not now regularly consumed, should be conserved for their food value.

A potential new food crop hails from northeast Africa. A bush called *Cordeauxia edulis* used to grow abundantly in Ethiopia and Somalia. The plant, distantly related to peanuts and soybeans, was one of the dominant shrubs of the area. This plant produces large seeds called Yeheb nuts with a pleasant sweet flavor which were eaten either raw or cooked. A recent analysis showed that the food value of the Yeheb nut is surprisingly high—53 percent carbohydrate, 13 percent protein and 12 percent fat. As early as 1907, the Yeheb was described as being the staple food of people in that portion of the Horn of Africa. Yeheb nuts also were ideal because they thrived under arid conditions, occurring naturally where rainfall amounts to barely ten inches a year.

Unfortunately the people of the area did not husband their resources wisely. *Cordeauxia* is now on the endangered species list. The farmers did not leave sufficient seeds on the plant to replace the mature plants nor protect the scant number of seedlings that did germinate from the ravages of goats and other browsing animals. The wild plant populations now are impoverished, having dwindled to just three small arid areas. While *Cordeauxia* is endangered, hope does exist for this plant. Attempts are being made today to grow Yeheb nuts as a crop in both Somalia and Kenya. In order to make Yeheb nuts a highly effective crop, the food yield must be increased. Each seed pod grouped in a terminal cluster, bears one or two nuts. We need to breed plants that bear more seeds to the pod, more pods to the cluster, and more clusters on the plant. This type of breeding is possible, but it would have been far simpler and much less expensive to have improved the crop with wild plants that carry the desired high-yield traits. But the diminished size of the wild plant gene pool prohibits this approach. Other valuable traits may have been lost as well. Some of the wild species may have contained the ability to survive in desert areas so dry that the plants needed barely any irrigation; these genes probably have been lost, too.

Some is Better than None

The gene pool of a population may be markedly reduced before extinction takes place. At this point in the extinction process, scientists split into three groups. Some question saving only one or two plants when no others of the species remain. Other scientists question saving a collected specimen of a species which has lost its records of origin. The third group, to which we belong, considers that half a loaf of bread is better than no loaf at all. The plant may have lost its value for the scientists who name plants and worry about their natural distribution, but nonetheless, it still is living and is intrinsically important. The preservation and study of these plants ultimately may prove to be helpful to mankind.

The Ultimate Fate

The ultimate result of a dwindling population and gene pool is extinction. Within the natural process, two kinds of populations and situations lead to extinction. In the first situation, opportunistic species invade new territories with different environments. These opportunistic species generally not only find a new habitat rapidly but are able to utilize available resources efficiently. They are naturally short-lived with lifespans usually less than one year, and they produce numerous offspring that can be disseminated over a wide area. Populations in any one area tend to be small and, if extinction occurs, it occurs abruptly, usually due to catastrophic events with which the population cannot cope. When a tree dies in a forest and makes a clearing or when a field is plowed and then allowed to lie fallow, opportunistic species—frequently weeds—invade. These species seldom last for extended periods of time.

In the second kind of extinction, large populations exist in a stable environment, but at the maximum capacity the habitat can provide. These species tend to occupy the areas for many years on a continuous basis and individuals themselves may have relatively long lifespans. When these populations become extinct, they usually have overextended themselves and used up all of their resources, or they are unable to compete with another species that has appeared on the scene. These two extremes—the opportunistic pioneers and the stable stay-at-homes—are parts of a natural process.

There is nothing natural about the extinction rate prevailing in our present environment. Normally stable environments are being drastically changed through man's activities such as cutting down forests and plowing meadows. These changes in nature's habitat clearly favor the opportunistic species. The majority of species are not opportunistic and do not benefit from these activities. Most species are stable homebodies that cannot cope with these new situations. The only response to these catastrophic changes is extinction.

There is a general expectation that populations with few individuals become extinct faster than those with more individuals. Although this may sound logical, it does not necessarily occur. Many plant species are naturally rare, and only occur in nature as a few scattered individuals. Botanists suspect that some of these species always have been rare. The ability of these individuals—sometimes just a few thousand—to reproduce and maintain their populations remains a mystery. The success of such species could be due to the fact that the individuals occupy special microhabitats in which there is little competition. We can find a more definite explanation in a few groups of plants with small populations. Many members of the orchid family appear to be evolving so rapidly that individual species are not around long enough to establish large populations. They are continually splitting into new species. As we will discuss later, the destruction of most of this group may be imminent. The orchid family currently contains more species of flowering plants than does any other plant family, but by the end of this century, the picture will be very different.

We logically assume that the longer a species has survived, the more successful it will be at overcoming forces leading to extinction. However, people who study the lifespans of plant species have found that this actually is not the case at all. The chance of an event occurring that will lead to extinction is quite independent of the organism's previous history. This is one reason why so few species last for long periods of time and the numbers of these species drop off exponentially with time. It is much like a throw of the dice. The numbers that appear face up after a roll are quite independent of any previous throw of the dice. The same is true of events leading to extinction.

Previous Mass Extinctions

Biologists find that not only species become extinct, but groups of related species also disappear. Related animal and plant species are gathered together by taxonomists who give names to species and worry about their relationships to each other. They gather different but related species into categories called genera and related genera are categorized into families. For example, apricots, peaches,and cherries are different species in the genus *Prunus. Prunus, Malus* (apples) and *Fragaria* (strawberries) are all members of the rose family (Rosaceae).

Scientists recently analyzed data from marine animal fossils of the Cambrian era 600 million years ago and found that of the more than 3,300 families that had been recognized, 2,400 families were extinct. Extinct families were represented by species in the fossil record but they have no living representatives today. The scientists calculated the rate by which entire families became extinct and found that the rate averaged five families per million years. Sometimes it ran as high as eight and at other times it was as low as one or two families per million years. On four occasions, however, extinction rates were dramatically higher than the average rate. The first of these great extinctions was in the late Ordivician era (over 400 million years ago) when families were lost at the rate of 19 per million years. Nearly one hundred families—and countless species—were lost in the space of five million years. The fourth and last great extinction took place during the late Cretaceous (over 100 million years ago), when extinction occurred at a rate of 16 families per million years. This marine extinction event coincided with the demise of dinosaurs on land.

We can't help but wonder if we aren't currently in the midst of a fifth great extinction crisis. No one knows for sure what caused the earlier extinction peaks, but a variety of suggestions have been made. Hypotheses range from drastic climatic changes that were brought about by an alteration in the tilt of the Earth's axis to competition from better adapted forms of animals. One recent suggestion made to account for the demise of the land dinosaurs, and the simultaneous marine extinctions, is based on the possibility that a giant meteor struck Earth and affected the plants on the planet. An examination of the sedimentary deposits in the rocks formed during the Cretaceous crisis reveals a very thin layer containing Iridium nearly everywhere in the world. The Iridium was in a form which does not occur naturally on Earth, but has been found in meteors. Scientists suggest that a relatively large meteor (up to 10 kilometers in diameter) struck the Earth and, upon impact, either threw a large quantity of dust into the atmosphere or disintegrated and then formed the dust. The theory suggests that enough dust was kicked into the air to obscure the sun for about three months, thereby blocking sunlight from the plants which then could not photosynthesize. When these plants died, browsing animals were left without food. These, and other animals dependent upon them, then died out. Only scavengers or animals that could go for long periods of time without feeding survived the meteor impact. While this theory is not airtight, it is plausible and has been given considerable credence. It certainly points out this planet's ultimate dependence on its plant life.

The Trauma Continues

One day our era will be considered the close of the great age of mammals. What happened to the mammoth and the mastodon, the sabre tooth tiger and the woolly rhinocerous? Why did the great ground sloth and the big birds like *Aepyornis,* the roc from Madagascar, disappear? These prehistoric animals were not primitive. In fact, an examination of their fossil records reveals fossils of today's everyday mammals as well, proving that these now-extinct mammals once were part of the modern-day scene. The reasons for these extinctions are actually much easier to comprehend than the earlier disappearances during the age of the dinosaurs. The great mammals were killed off by early man as he swept across the continents. Even with his puny spears and clubs, man's greater intelligence made him more than a match for the big beasts. Unfortunately the Pleistocene extinctions of less than one million years ago are merely part of the present-day extinctions, which promise to be the most complete and devastating reduction of diversity in the five-billion-year history of the world. How much of the trauma being inflicted on this planet is borne out of necessity and how much results from pure and simple unthinking greed? Looking at what has already happened and what is happening again, it is apparent that it need not be this way at all.

Most of the preceding discussion has concerned animals. What about the plants? Very little of the fossil history of plants is known, primarily because the fossil plants embedded in coal deposits are burned, not studied. We do know that in the history of the world, many groups of plants appeared, flourished and then disappeared. Extinction works the same way for both plants and animals, under natural conditions. The rules of the game are the same for both. Unfortunately, we no longer are playing according to the natural rules.

Franklinia alatahama

Case History: The Franklinia Tree

The most famous North American endangered plant is probably *Franklinia alatahama,* a tree named after Benjamin Franklin. The history of the tree is well-documented. It was first noted on Oct. 1, 1765, by John Bartram and his son William along the banks of the Alatahama River in Georgia. William returned to the site a number of years later and collected some of the plants. In 1790, Moses Marshall, a nurseryman, went to the same spot and also collected material. He was probably the last person to see a *Franklinia* in the wild.

Franklinia, a member of the tea family, is closely related to camelias, as suggested by its shiny leaves and large white blossoms. A mature *Franklinia* gets to be thirty feet tall and makes a handsome garden specimen. During the late summer and early fall, *Franklinia* bears large flowers with five, rounded white petals and a central boss of golden stamens. Sometimes the three-inch flowers are borne just as the foliage turns scarlet in the fall.

These trees have been popular in cultivation for more than two hundred years. What happened to them in the wild? There really is no clear explanation. The species was rare when first discovered, with just a small patch growing at that site in Georgia. The trees must have already been on the verge of extinction, and so the small collections made by the Bartrams and Marshall merely hastened the species' natural end.

A number of attempts have been made to relocate *Franklinia* trees in the wild. Although there have been a few claims of success, all attempts ended in failure. This is one of the lucky species to be successfully cultivated prior to its demise in the wild.

CHAPTER 7

Tropics

The tropical forests—the end product of hundreds of millions of years of evolution—boast the richest species diversity ever seen on earth. Scientists know of at least 155,000 tropical plant species, as compared with 85,000 temperate ones. In the New World, the ratio is approximately 90,000 tropical to 10,000 temperate species. Not only is the tropical forest one of the most complex ecosystems in existence, but it is also among the most fragile and most threatened. Its illusion of fertility and lushness is merely a thin veneer. About 70,000 hectares of forests are being cleared each day, amounting to an astounding ten million hectares a year—equal to the area of the British Isles. Many tropical countries have already lost all of their tropical forests; others will lose their small remnants within a few years.

The tropical forests are the least studied of the world's vegetation types. Scientists estimate that just 1 percent of the tropical species have been assessed for their potential usefulness. We know that the forests are a vast storehouse of important genetic potential with resources for scarcely imagined products. The tropical forests actually hold trees that make pure diesel fuel (*Copaifera langsdorfii*) as well as plants that may cure cancer (*Catharanthus roseus*).

The destruction in the tropics has important implications for the world, even those who live in the affluent temperate countries. In this chapter we'll discuss the plight of the tropics and the ramifications of the present shortsighted actions being taken in tropical countries. Many of the world's top scientists have called attention to the attrition of the tropical forests, but little has been done to take their advice or heed their prophecies. If a significant portion of the forests are to be saved, it must happen during this decade. Time is running out.

Changing Our Weather

Tropical forests play an important role in the recycling of carbon dioxide. The trees and plants take carbon dioxide out of the air to make organic molecules and then release oxygen from water in the process. Trees make our air breathable. For a long time scientists believed that the major recycling of carbon dioxide and the release of oxygen was conducted by phytoplankton, the microscopic plants that live in the ocean. More recent analyses show that the major recycling comes from the terrestrial plants, particularly the ones in the tropical forests. The amount of oxygen released into the air is directly related to the amount of carbon fixed when a plant photosynthesizes. Despite the fact that the ocean surfaces measure nearly two and one-half times land surfaces on Earth, land plants fix nearly five times as much carbon as do ocean plants. Furthermore, moist tropical forests can fix nearly three and one-half times as much carbon dioxide as grasslands or cultivated croplands. As we convert tropical forests into cultivated fields or pastures, the planet's ability to recycle life-giving oxygen may be drastically reduced.

The world will not run out of oxygen but there will be a buildup of carbon dioxide. There is hard evidence showing that a significant increase in carbon dioxide has already occurred. The projected outcome for this is what is known as the "greenhouse effect." The theory is that because carbon dioxide will block the escape of heat from the sun, the increase in carbon dioxide will raise the average temperature a few degrees. This could lead to mass melting of the polar ice caps which, in turn, would raise the sea level. Enough of an increase would cause destruction of the major cities of the world by inundation.

If conversions of tropical forests to croplands are unsuccessful (and we have every reason to believe that they will be), the croplands will end up as arid deserts. Contrary to popular belief, the soil in tropical forests is not rich. A limited amount of nutrient is contained in the soil, and this is rapidly recycled through the organisms of tropical ecosystems. When a tree or plant dies, it is quickly attacked by decomposers—fungi, bacteria and wood-eating insects. In no time at all the trunk will disappear as it is recycled into other living things. By contrast, tree trunks in temperate forests can linger for decades after the trees have died. Soil in temperate regions has a great store of organic matter and humus, but there is little of this in the tropics. When tropical forests are cleared, one or two seasons of good crops may be cultivated before exhausting the soil. Within a few more years the soil becomes a concrete-hard surface, totally useless for agriculture. Such barren wastelands surround many of the tropical cities that once had lush forests at their borders. Humid forests can easily turn into dry deserts when the trees are cut down. We already see deserts encroaching down the African continent. There is a very real threat of the great Amazon being converted into the counterpart of the Sahara.

Trees, the link between water in the soil and water in the air, transfer considerable amounts of moisture through their leaves. It is not clear how much moisture from the trees is re-precipitated as rain in the vicinity of the forest and how much moisture is carried to other areas as rain. We do know that of the rain that falls in and around the forest, 50 percent is recycled by the trees back to the air above the forests. Some scientists fear that deforestation could drastically affect the rainfall in certain parts of the world.

Watersheds, Erosion and Floods

If forests are no longer around to suck water from the soil and return it to the air, what happens when it rains? Without tree foliage to break the fall of the pelting tropical rain and tree roots to hold the soil, erosion increases and the thin tropical topsoil washes away. Much of the water that would have been recycled into the air by the trees finds its way into streams and then rivers. Devastating floods can result. Studies made in Peruvian Amazonia confirm the rising water levels. During the last twelve years, more than 51,000 square kilometers of the Amazonian forests in Peru at the base of the Andes have been logged and cut away. Water levels of the Amazon were measured at Iquitos from 1962 to the present. The highest level during the rainy season never rose above twenty five meters until 1969. Since 1969 the average level has been over 27 meters. The increased water level during the rainy season is undoubtedly due to the poor retention in the soils of deforested areas. In 1979, severe flooding swept through the Belem-Brazilia region and a six-week drought occurred in Manaus. Both Belem and Manaus are industrialized areas of Brazil and economically important regions. In both cases, deforestation was responsible.

Similarly, Venezuela's deforestation has led to greater water runoff during the rainy season. However, an additional side effect has been seen in this country. The forest used to hold back water and drain it into the rivers on a year-round basis. Now the streams are seasonal. Dry stream beds during the dry season jeopardizes the Venezuelan llamas. Many villages and settlements are built along the edges of rivers in the Amazon basin. We can expect the increased logging to bring about devastating seasonal floods.

Across the globe in India, floods caused by deforestation have led to a unique folk movement called the Chipko. Following extensive deforestation, a flooding of the Alakananda River in the Chamoli district of India devastated the village of Belacuchi and swept away several tourist buses, twenty bridges, hundreds of cattle, and many hundred million rupees worth of timber and fuel. Every monsoon thereafter, disastrous flooding occurred. The villagers realized that not only must logging cease, but that reforestation was necessary. Demonstrations were organized against the government's logging policies, but these were of no avail until some people clasped their arms around the threatened trees and remained there even as the loggers arrived to chop down the trees. Chipko, literally meaning "hugging the trees," was tried successfully many times. On one occasion, 27 women of the Reni village saved nearly 2,500 trees. Several times during the ensuing six years, people from various villages resorted to Chipko and succeeded in saving remnants of their forests when the government changed its logging policies. Attempts to replant forests have also proceeded in parts of India with plantings in the watersheds, and willows and poplars along stream banks. Bear in mind that reforestation was not particularly successful until the local people were told they could use the trees when mature.

Assessing the Damage

Most of the world's tropical rain forests now exist in just a dozen countries. Of these, the largest forest is the Amazonian forest of Brazil with 2,800,000 square kilometers. Despite the fact that the forest has been called a green hell, it is Brazil's future source of wealth. In the current economic climate, quick and immense profits can be realized by converting the forests to pasturelands for beef production. However, the frequently overgrazed pastures can support few cattle and are good for only a few years. More forests must then be felled to make more pastures for the cattle. A similar pattern can be traced in Central America. Clearing the forests for cattle ranching appears to be primarily a Latin American phenomenon.

Next in size are the forests of equatorial Africa, primarily in Zaire and Gabon, followed by the forests of Malaysia and Indonesia. Other countries with extensive forests are the South American countries that share the great Amazon—Peru, Colombia, Venezuela, the Guyanas, Bolivia and Ecuador. Other great tropical forests existed in years past. Madagascar was 85 percent forested but now has less than 8 percent of its land covered by forests. Thailand was 75 percent forested at the end of World War II; now just a tiny percentage remains. East Africa's mountain forests are mere relics, as are Sri Lanka's rain forests. The forests clustered at the base of the Himalayas in India and Nepal have been cut and pillaged until hardly anything is left. Only 38 percent of the Philippines is still forested.

Dr. Norman Myers, stationed in Nairobi, Kenya, has compiled a telling indictment of the ruin of the world's tropical forests. He probably has a clearer understanding of the extent of the problem than anyone else, having traveled to examine and assess tropical forests in some 43 countries. The Food and Agriculture Organization (FAO) of the United Nations has estimated deforestation rates at 20 hectares per minute, which works out to nearly 30,000 hectares per day or about 150,000 square kilometers per year. Myers' calculations are twice as negative: he estimates the loss is 46 hectares per minute, 66,240 hectares per day and 245,000 square kilometers per year. Myers says the FAO figures are an understatement because they are based to some extent on pre-1970 figures. Since then, exploitation has increased, not decreased.

Myers has listed nineteen distinct forest regions where little or no forest is expected to remain by the 1990s. These are areas where conversion rates are most rapid, and where the likelihood of significant forest regions being set aside is slight. Small preserves may exist, but in all probability they are mere tokens when compared to the areas once covered with lush growth. With the exception of Lowland Burma and Papua, New Guinea, all Asian tropical forests will be converted by 1990. Forests will be lost not only from the Asian continent but also from the East Indies. Australia has a strip of tropical forest along its northeast edge which is also threatened. It is somewhat ironic that eucalyptus species, some of the major fast-growing plantation trees of the world, hail from Australia but are not grown in sufficient quantities on that continent to relieve the pressures of deforestation.

In the New World, all of the Central American forests are listed. Both El Salvador and Nicaragua have already depleted their resources. Further south, Ecuador's Pacific Coast forests, the eastern and southern sections of Brazilian Amazonia and Brazil's Atlantic coastal forests all will have been exploited by the 1990s.

The situation is no better on the African continent. East Africa's forests are confined to mountain relics, West African forests are expected to be cut by the end of the decade, and Madagascar has only remnants left.

By the turn of the century much of Papua, New Guinea as well as remaining sections of the Burmese Lowland forests may have been cut. This will leave no significant forests on the Asian continent or among the Pacific island systems. Most of Ecuador's and Peru's Amazonian forests will be cut if logging continues at the present rate. Only three areas will have significant amounts of forests left by the year 2000: Western Amazonia in Brazil, a lowland rainforest that is wetter and more difficult to settle; the South American forests of Guyana, Surinam and French Guinea; and the Zaire basin in Africa, another lowland forest. Myers expects these to remain relatively unexploited. Zaire forests have escaped total exploitation because the population is low and there are sufficient mineral mining resources to support the economy.

Exploitation in the Tropics

Tropical forests are reservoirs of lumber, containing old hardwood trees used to make furniture and veneers. The wood also is used to construct buildings and to make paper, pulp and fuel. The demand for wood by the developed countries continually increases, furthering the exploitation of the forests. There has been a staggering increase in demand during the last few decades. Myers pointed out that in 1950 only 4.2 million cubic meters of lumber found its way to the developed nations. By 1973 the amount was 53.3 million cubic meters. At this rate the demand will be 95 million cubic meters in twenty years. The demand for tropical woods has doubled in tropical countries while demand in the rest of the world has grown fourteen-fold.

Logging in tropical forests usually is highly selective. Just a small percentage of the trees are prime lumber species. However, when these few trees are harvested, 30 to 60 percent of the remaining trees are severely damaged in the process. Harvesting trees more carefully to protect the remaining trees is expensive and would raise costs. Historically, profits have proven to be more important than forest preservation.

As mentioned earlier, forests also are exploited because of the increased worldwide demand for beef. The appetite of the world for "Big Macs" and "Jumbo Jacks" is prodigious. Countries of Latin America are converting their fields to pasture lands to meet this demand. By 1979, Brazil had converted some 80,000 square kilometers of forest to cattle ranches by killing the forest with chemical herbicide and then setting it on fire. Peru and Colombia have seen the drive in Brazil to open up the forests to ranchers and are following suit. Much of this conversion is being financed through foreign investors. Cattle production does create a few jobs; other than that, the Brazilians get little from their ruined forests. Most of the beef is sent to the richer nations that can afford fast food restaurants; the profits make their way into the pockets of a few wealthy agriculturalists and foreign investors. We can see the picture clearly in Costa Rica where one-third of the forests were cleared for cattle ranches during the 1960s. Beef production in that country doubled, but the per capita consumption of beef in Costa Rica dropped by 26 percent.

Farming is speeding up the decline of forests in Asia and Africa. There, forests are being converted to small, subsistence farms. Small fields are cut out of the forest, planted for a few years, and then abandoned. Years ago the abandoned field could regenerate into forest, but population pressures are so strong today that the forest does not have a chance to heal its wounds. Myers estimates that 100,000 square kilometers of forest land are eliminated each year to accommodate these small farms. So far, one-fifth of the tropical forests of the world have been converted to subsistance farms. The problem is that the population these farms support will double by the turn of the century. Demands for land will be ever-increasing.

Fuel is another reason to exploit tropical forests. Half of the wood cut in the world is burned as fuel. In most developing nations a forest must be more than 15 kilometers away from a village to be safe from wood gatherers. The secondary forests attempting to regenerate frequently face the most pressure for wood as fuel. As long as these forests are ravaged for fuel, regeneration will not occur. Only about 20 percent of the fuel wood gathered is derived from primary forest, but that accounts for about 25,000 square kilometers of forest per year. We are not referring to logging operations but to people simply heading into the forests to collect wood to cook their meals and warm their homes. Nearly half of the yield of logging operations is also burned as fuel. The demand for wood as fuel has recently increased because of hikes in oil prices. Even the oil shieks bear some responsibility for the debilitation of the world's forests.

All Forests are not Created Equal

When a country reports its forest land, the reports frequently include the total sum of all kinds of forests because of the tendency to confuse different kinds of forests. Yet all forests are not alike. Primary forests are virgin forests, the climax vegetation that is the end result of natural growth. When primary forests are disturbed or logged, they are replaced with secondary forests, usually composed of faster-growing but smaller species. In the tropics, secondary forests are never as rich as primary forests as far as species diversity. They are rather impoverished by comparison. Preserves sometimes are called forests even though they contain no trees. Many forest "preserves" in the United States are nearly treeless, such as the Cleveland National Forest in Southern California which has mainly scrub vegetation.

Another type of forest is the manmade plantation, usually composed of a species of pine or eucalyptus. Often called monocultures, these forests consist of only one species. Amazonian forests contain more than 2,500 tree species. Replacing such a forest with pine is a very poor substitute yet many countries do not differentiate between plantations of a single species and natural forest. Replanting deforested areas with a monoculture is not true reforestation. Pine trees and eucalyptus are ubiquitous species planted all over the tropics. They produce an inferior wood, but they grow fast enough to make good trees for pulp and fuel and are better than no trees at all or arid baked deserts. Unfortunately, few other plant species can tolerate growing near them. Many animal species can't tolerate them as well.

There is a famous experiment in Brazil known as the Jari Project. About 100,000 hectares of land were cleared and planted with two species—*Pinus caribea* and *Gmelina arborea*. A special pulp mill was built in Japan, floated intact across the world, and up the river to the Jari plantations. The plantations produce about 750 tons of pulp and are roughly twenty times as productive as natural forests. It still remains to be seen whether or not the land can sustain continuous yields of trees. Because Jari has exceptionally rich soil, this experiment probably should not serve as a model for long term expectations. It does demonstrate that some areas could possibly be farmed to reduce pressures on the remaining natural forests.

The Future?

Within the past few years there has been a dawning consciousness in many tropical countries that their willingness to turn a quick profit may be jeopardizing potential future resources. We see a growing trend to set aside preserves and to create policies to maintain them. The current Brazilian government has taken the lead by initiating ambitious conservation projects. Developers in Brazil are expected to maintain half of the land in its natural state. 600,000 additional square miles of national park land and 250,000 square miles of national forest conservation land are supposed to be set aside. This 250,000 square miles represents about one-eighth of the Amazon basin. This sounds great, but we have seen many national policies come and go in Brazil as governments have come in and out of power. We can only hope that the current attitudes toward the great Amazon will be maintained by subsequent administrations. Goodwill is ineffective unless the good intentions are carried out with some determination. Many projects seem wonderful on paper but are followed through in a haphazard and useless manner. Intent to save a forest does not automatically mean that the forest is on its way to being saved.

Clearly, we are now facing the critical years for the world's tropical forests. The policies of the tropical countries must undergo radical change in the immediate future if we are to retain more than small island preserves where species will continue to decline at a steady rate. We look back on the devastation of island floras and blame the early settlers and sailors for introducing herbivores with voracious appetities such as goats, deer and rabbits. They did this out of ignorance or because their view of the world was anthropocentric. We have no rational excuses today for the conversion of tropical forests. Our only reason has been greed. If the present rate is maintained, we know we will see the end of the great forests in our lifetime and can expect to witness one of the greatest and fastest periods of extinction in the history of the planet. How will we answer our children and grandchildren when they want to know why we squandered their inheritance? We can't say we are not to blame because the cutting took place half a world away. Two-thirds of the products obtained through these conversions are exported to the developed countries of the Western world. Whether we like to hear this or not, we are responsible.

Saintpaulia ionantha

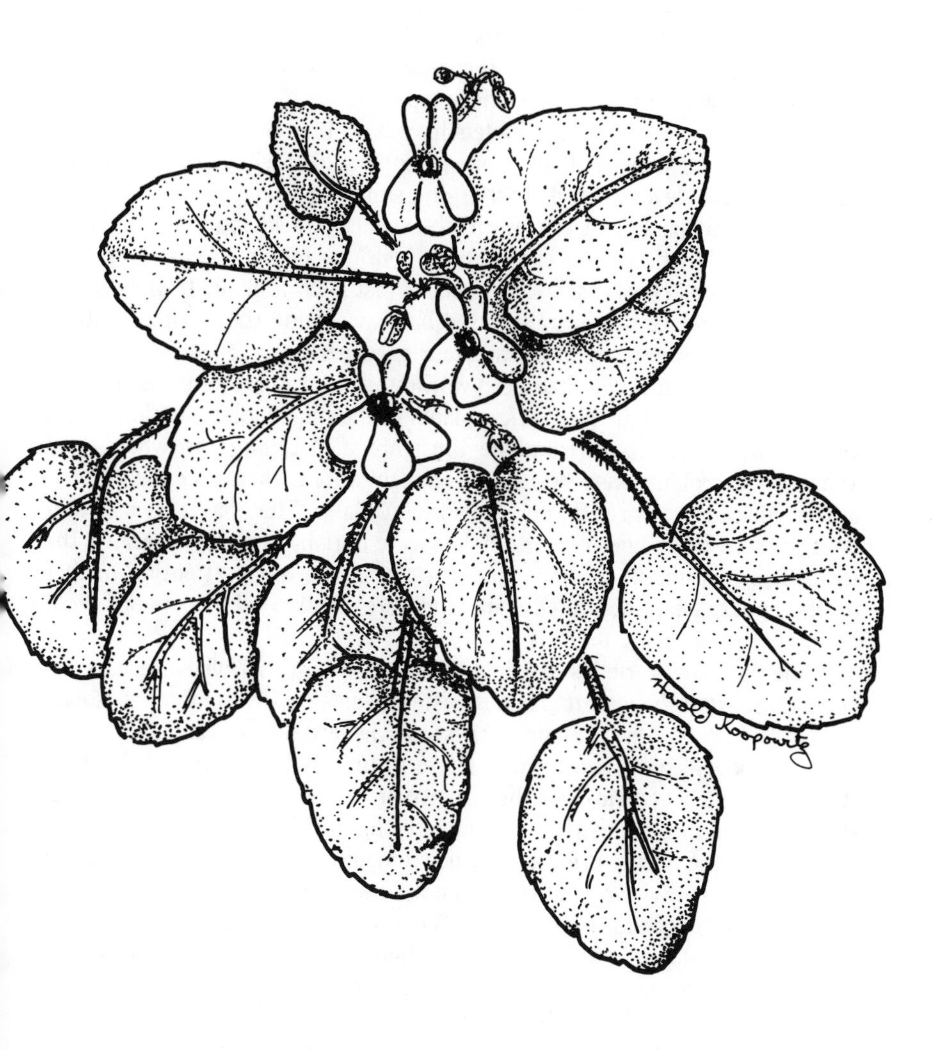

Case History: The Primitive African Violet

What could be more ubiquitous than an African Violet? These little furry plants with small but colorful blossoms clutter the shelves of supermarkets and drugstores, gathering dust as they wait to be picked up and deposited on a kitchen window sill or an office desk top.

In the last fifty years African Violets have been bred by the thousands, with so many different varieties that no one person could possibly grow all of them. Yet all of these varieties started off as one simple species.

Only a specialized plant can live in the vertical cracks found at the base of cliffs, especially where trees encroaching upon the cliff walls cast deep shade upon the plants. The Umsambara violet is just such a plant. It grew in that unlikely spot before it was transported halfway around the world, eventually to become the sweetheart of the amateur gardener.

In the 1890s, nuns who had been working in East Africa returned home to Germany and took with them a few plants of the blue, violet-like flowers that grew in the forest shade from the Umsambara Mountains, in what is now Tanzania. The plants were new to science and were named after the governor of that African province, Baron St. Paul von Illaire. The plant's success story does not start here, however. The violets were initially unpopular because they seemed to be difficult to grow. In fact no one paid much attention to them until they arrived in the United States and were popularized by a famous orchid nursery in Southern California.

While the many varieties based on these initial African Violets are extremely popular, *Saintpaulia ionantha* and its sister species are still confined to isolated hills back in East Africa. Many of these species are now endangered because the local people have cut down the forests in search of firewood. With the shade gone, the hot African sun beats down on the remnants of the Umsambara violets—bleaching the hairy leaves and burning the blossoms. The latest tallies show that of the twenty primitive species remaining, a number are threatened. Among these is the ancestral cultivated African Violet.

Those who successfully grow African Violets know very well that these plants do not tolerate standing water or water in their crowns. In nature, their position on vertical cliff faces precludes any water gathering in areas not suited to the plants. The fact that these plants are so specialized in their habitat spells out their doom. Unless the cliffs can be reforested and habitat degradation halted, their extinction seems inevitable—a sad end for the ancestors of such a popular plant.

If African Violets were animals, their popularity would certainly have stimulated a great outcry once people learned that the wild relatives were threatened.

People would set up a preserve and concerned individuals around the world would send money to maintain the wild endangered species. The reality is that violets are plants, not animals. Few people seem to care that the wildings are fading away.

CHAPTER 8

Islands

Looking through the pages of the Red Data Book on threatened plant species, one fact become obvious: island-living is not paradise for plants. On the island of Socotra, 132 or nearly three-quarters of the 216 endemic flowering plants are threatened. The majority of the 116 species that occur on Juan Fernandes Island in the South Pacific are in danger. More than half of the flora of Tenerife in the Canary Islands are endangered and the pattern is repeated on other islands in the group. On the Mediterranean island of Crete, 101 of the 155 species are in trouble. In the Seychelles in the Indian Ocean, nearly all of the 72 species are threatened. Everywhere in the world the problem is the same. Islands in the Antarctic Ocean suffer as much as the tropical isles of the Pacific and the temperate islands in the north. Whether they are inhabitated or uninhabitated, islands appear to be vulnerable and fragile. As we'll see later in this chapter, mankind is indirectly responsible for the difficulties faced by plants, even on the uninhabited islands.

Island Ecology

The rules that govern the ecology of islands are, in some respects, different from those of continental masses. Continental species have numerous opportunities to move around from one place to another. They can migrate into or out of new habitats. Non-flying organisms on islands have a relatively difficult time crossing the expanse of water surrounding each particular island. Immigrants arrive on the isle purely by chance. Seeds or spores can be carried by birds or occasionally lifted long distances by violent winds or storms. The further from the mainland, the fewer the species. When they arrive and begin growing, they may find that since few competitors have had the luck to find the island, they will succeed and evolve. Because so few organisms of any one species reach the island destination, the founding members can have a profound effect on the population. This is like the tale of the Scotsman who was shipwrecked on a South Pacific island. Four generations down the line, half the population had red hair. Thus plants on different, though closely related islands can be related, yet still look distinctly different.

Unlike physics, biological laws are few. One law which appears to hold true concerns the number of species that an island can maintain. Regardless of whether the subject is plants or animals, the number of species found on the island is directly related to the size of the area. For each tenfold increase in area, an island usually will double the number and kinds of species it can maintain. Therefore small islands have fewer kinds of birds or flowers.

Islands are formed in one of two ways: either they grew out from the sea (such as coral atolls and volcanoes) or they are areas that were separated from land masses when the sea level rose after the last ice age. If a chunk of mainland which had become isolated and formed a landbridge island (like Britain) was isolated after the last ice age, the newly formed island would begin to lose species. Biologists have confirmed this phenomenon on numerous islands in many parts of the world. It is possible to work out how fast landbridge islands have lost species and how these related to the size of the islands. Good examples of large landbridge islands are Britain, Tasmania, Sri Lanka and Zanzibar. The entire Indonesian area is one such cluster of literally hundreds of islands of varying sizes that were created by rising sea levels. Some are large (such as New Guinea), and others like Singapore are small. The glaciers started to melt about 14,000 years ago and the sea reached its present level about 6,000 years ago. Islands like Sumatra and Java have about 70 percent of the species found on the mainland, whereas smaller places like Bali have about 10 percent of their initial number. The rate by which species are lost is inversely related to the size of the island. A small island can lose most of its species in a thousand years. A thousand years may seem like a long time in terms of human life, but it is just a wink of an eye in terms of the geological life of the planet. Today's extinctions, some of which are occurring over decades, will seem like an instantaneous catastrophe when viewed from the future.

The biological law that dictates the welfare of island species raises some important considerations for conservation and the creation of preserves. The island diversity law holds true not only for pieces of land surrounded by water, but also for other isolated areas. These areas of isolation could be mountains rising out of a plain or perhaps a remnant cluster of forest. An acre of field surrounded by concrete skyscrapers is, to all effects, an island. Biologists in the Amazon have found that when a patch of forest is surrounded by a clearing, many animals are reluctant to cross to other patches of forest. Some forest birds will not cross a hundred yards of open space. Their clump of trees may just as well be surrounded by miles of seawater. Koopowitz observed this phenomenon in Africa, having the good fortune to spend ten days exploring the country of Lesotho. Most of this country is at high altitudes and many of the plants are alpine. Another time Koopowitz explored another area about 300 miles to the southwest of Lesotho where he climbed the highest mountain in the area. There, at the 6,000-foot top, grew some of the species seen in Lesotho. There were just two or three of these plants and they were very different from the ones growing on the lower slopes of that same mountain. The mountain top was, in effect, a little island with an impoverished number of species.

Goats

We cannot discuss the ecology of islands without making a few disparaging comments on goats. These creatures must be the true embodiment of the devil for a plant lover. They have pillaged and despoiled islands and continental masses alike. By nature they are almost omnivorous—they defoliate all plants within reach and even climb into bushes and scramble up trees to reach inaccessible green morsels.

In the sixteenth and seventeenth centuries goats were automatically introduced onto islands as they were discovered. The early explorers considered goat to be fresh meat on the hoof. Goats seemed ideal because they could run wild, fend for themselves, and reproduce effectively—particularly in the absence of large predators.

St. Helena, a small island in the middle of the South Atlantic Ocean, has played a role in French and English history. Today it is a British outpost of strategic importance with a small agricultural population. The Portuguese discovered St. Helena in 1502 and introduced the goats eleven years later. The animals spread and devastated the unique flora, as they did on countless other islands. There are records of goat herds up to two kilometers long within 75 years. St. Helena originally was covered with forests, but by 1800 these were reduced to remnants on the central ridge of the island. J.D. Hooker, a botanist of the time, estimated that there must have been one hundred endemic species. Only after the goats were wiped out in the 1950s did some of the vegetation start to rejuvenate. Even so, the flora has been impoverished from the original one hundred down to 33 endemic species. Of these, 18 are still classified as endangered.

The Effects of Man

As would be expected, islands that are permanently settled by man are facing the greatest threat to their natural resources. The inhabitants have not only misused their resources on some islands, but the expansion of the human population on those islands directly threatens their survival. I (Koopowitz) remember a trip I took many years ago to islands in the South Pacific. One particular island, Fakaofo, stands out—both for the hospitality of the people and for their plight. Fakaofo, one of the Tokelau Islands, is a tiny atoll surrounded by a ring of coral sand scarcely a quartermile wide. More than six hundred inhabitants eked out a living from the sea. The island people also farmed pigs, which were responsible for the destruction of much of the island's natural vegetation. The islanders would toss coconut husks into the central lagoon and the pigs would wade into the water to forage on the rotting husks. This was a recycling process but the net result was the destruction of the lagoon. The people there were kindly and hospitable, but now, a generation later, I can't help but wonder how many people are crammed together, and if the tiny island will end up like a crowded rat cage, with all the aberrant behavior found in a laboratory situation.

Closer to the coast of South America are the Easter Islands—grassy outcroppings famous for their great statutes. Archaeologists have found that the early peoples used wood to build canoes and construct their houses. Now there are no naturally occurring trees or shrubs. In 1956 the last specimens of *Sophora toromiro* were recorded but they had disappeared by 1962. This species was the source of wood for the islanders, and it appeared to be endangered even at the turn of the century. Only a single tree was known in 1917. The species persisted, however, because this tree produced seedlings. Some of them matured before they succumbed. The final plunge into extinction was due to sheep grazing and not allowing the seedlings to survive to maturity. When Thor Heyerdahl, the famous explorer, visited the Easter Islands, he collected seeds from the last tree. Those specimens are alive in Goteborg Botanic Gardens in Sweden. If these trees can reproduce, there may still be hope of reintroducing *Sophora* back to its original islands.

Antarctic Islands

A small number of islands that dot the southern oceans have scarcely been touched by man—mostly because the climate has been too intemperate. During the great whaling era, however, these islands were popular stopping places where cooped-up sailors could stretch their legs.

Whalers and early explorers introduced a number of grazing and browsing mammals to the Antarctic Islands during the 1800s. These animals were used as a source of fresh meat, both for shipwrecked sailors and visiting whalers. This was before the days of refrigeration. Often the only available meat was salt pork or dried jerkies. The whaling ships were away for months at a time and their closest sources of fresh food were the Cape of Good Hope, Chile and Argentina, or New Zealand. The "harvest" from the islands provided a welcome relief from the monotony and dangers of the unpredictable Southern Oceans.

When the whalers and sealers began to visit Antarctica, only one mammal was native to the islands—the Falkland Island Fox. The fox was quickly exterminated by the farmers who settled the Falklands. In all, nearly twenty alien mammals were introduced. Sheep, goats and rabbits were set free on most islands; cattle and horses were left on others. On South Georgia, near the Falklands, reindeer were also introduced. The only permanent human population was established on the Falkland Islands, where sheep farming was the major industry. Perhaps the greatest loss on the Falklands was *Poa flabellata,* a particularly nutritious pasture grass that was grazed to extinction. If this grass were still in existence, it could be used to create more nutritious pasture grasses for other parts of the world. Grasses on other islands also were decimated and even lichens are endangered.

Islands in the New Zealand sector lacked large numbers of colorful flower species but the species that did exist were among the showiest. They were so delightful that in 1847 botanist J.D. Hooker described the flowers as "unrivaled in beauty." These islands included the Aukland Islands, Campbell Islands and Adams Islands. The only remnant of this beauty is on Adams Island. New Zealand has set aside preserves to try and save what is left.

MacQuarie Island is a particularly disturbing case. Rabbits let loose on that small island totally eliminated the main species. The animals grazed the plants so low that the roots died. Because the roots of the dominant grass bound the topsoil together, erosion started when the roots died and the topsoil was swept away. Before long MacQuarie was stripped of not only most of its soil but also most of its plants. MacQuarie Island was proclaimed a wildlife refuge in 1933 by the Tasmanian government. It is a poor preserve when compared to what it might have been.

On other islands, when the dominant species was lost by overgrazing, the ecosystems were considerably altered. Rabbits obliterated the dominant vegetation cover in the Kerguelen Archipelago that was composed of three different plants: *Azorella, Pringlea* and *Festuca.* In their place grew *Acaena,* a plant that recovers rapidly from grazing and is even spread by rabbits. Unfortunately, the unique, small invertebrate animals that lived on the original vegetation could not adapt to *Acaena* and died out.

The British Isles

The British are renowned for their gardens and love of plants. Despite their dense population and high level of industrialization, the British people have a long involvement with conservation. The British and British-trained plant people around the world have been leaders in the fight to avert the current world crisis in plant extinctions.

The first English conservation laws go back to 1535 when King Henry VIII signed a bill "to avoid destruction of wilde fowl, to protect eggs of various birds." Attempts were made to reestablish some of the woodlands in the seventeenth century, but concepts of nature were very different from today. Perfection was seen in clean, geometrical shapes, straight lines and circles. Since nature abhores straight lines, nature was considered degenerate and corrupt. Within a century British concepts had changed drastically and nature was seen as being noble and pure. Nevertheless, by the time the thinking had changed, the English had demolished their native, untidy forests and woodlands and replaced them with farms composed of small, neat fields. Around the fields stone walls or, more frequently, hedgerows were erected. The hedgerows and meadows became refuges for many types of plants. As hedgerows are now being cleared to make larger fields needed for mechanization, we find many plants becoming endangered. About 18 percent of the British flora—about 321 species—are listed as being in trouble. Many preserves have been erected specifically for plants as the British attempt to save what is left of the natural vegetation in the British Isles.

Japan

The Japanese islands are on the other side of the world. These islands also are highly industrialized, heavily populated, and have a long history and love of gardening. About 68 percent of Japan is still forested, but these forests frequently are not the natural mixed broadleafed species of primary forests. Instead, they are either secondary forests that replace the primary forests after they have been cut down, or they are plantations of conifers. A complex system of national, quasi-national, and prefectural parks has been set up in Japan with about 8 percent of the total land space set aside for the park system. There is also a Reserve Forest System where both animals and plants can be protected. Private land ownership is permitted in the park system. Even though laws exist to allow for transfer of land, they are seldom enforced. Some of the parks have a large degree of public access, and these are the ones that show the most deterioration. Many worthwhile parks exist in alpine counties of Japan, but there are few high quality parks in other regions.

Japan also has a unique system of Shinto shrines which are small areas of forest that are considered to be sacred. These contain some of the oldest and least damaged forests of the islands.

Extensive lists of threatened and endangered plants have been drawn up for Japan's flora. Unlike the endangered species lists of other countries, Japan even cites endangered fungi and mosses—plants other countries tend to ignore.

New Zealand

In contrast to highly industrialized groups of islands such as Japan, the islands that comprise New Zealand are an agricultural paradise with a modest population. Even so, New Zealand inherited severe ecological problems from the misdirected actions of its early colonists. Because New Zealand was geographically isolated from other land masses for such a long time, it possessed a very small number of endemic species. Along with there being just one small land mammal (a species of bat), the native flora was quite sparse when compared to Australia, New Caledonia and other islands. A few spectacular plants did exist. When the British arrived to colonize the islands, they were homesick for the British countryside and consequently imported their favorite birds, mammals, and plants. These species spread throughout the countryside, threatening the local birds and plants, all of which were unprepared for the competition. Besides the threat of being squeezed out by foreign species, some of the more spectacular plants, like *Ranunculus glacialis,* a great white buttercup, suffer from over-collection by alpine plant enthusiasts. One notorious animal pest, the Australian opossum, has established itself in New Zealand and is responsible for the deterioration of quite a few species.

New Zealand has set aside about 6 percent of its territory as parks. The Tongariro National Park, created in 1894, has an interesting history. As a precaution against having their holy places desecrated by the British settlers, a Maori chief, Te Heuheu Tukino, deeded the holiest spot in the mountains to Queen Victoria as a "sacred place for the crown." He hoped the area would receive permanent protection. And it did.

Malagasy

Madagascar split off from the African continent to form one of the most floristically interesting islands in the world. Many of the plant groups evolved in isolation on the island to produce some spectacular flowering plants. Forests on the island originally covered 80 percent of the land surface; only 8 percent is still forest. Tiny forest remnants now cling to the more inaccessible areas of the island. The remainder of Madagascar was arid and featured some spectacular succulents. The number of threatened Madagascan succulents in cultivation in botanic gardens around the world was assessed in 1980 by the Botanic Gardens Conservation Coordinating Body. The survey revealed that of the 473 species native to Madagascar, 297 are being cultivated in 42 botanic gardens, primarily ones in southern Africa and Europe. It is troubling to see that no garden from Madagascar itself is listed. Certainly the people of that country must be aware that their flora is unique and that there is worldwide interest in their plants and their maintenance.

Hawaii

American botanists were shocked when the numbers of endangered Hawaiian endemics were announced. At least 273 species or sub-species have already become extinct and another 800 are endangered. The causes are the paving of the islands and the push to build more and more condominiums and hotels, plus the intense agriculture—primarily pineapple fields, sugar cane, and cattle grazing on several of the islands. Another reason for the troubles in Hawaii is that weedy species such as *Lantana* were introduced years ago and have become rampant throughout large areas of the islands. The new species have had a competitive edge over the natural native plants and in many places have pushed the older varieties into extinction.

Efforts are now being made to save the remnants of the native flora. The Pacific Tropical Botanical Garden has been established to propagate and study tropical vegetation. This garden on the island of Kauai has been able to propagate and disburse a number of the endangered species. Special preserves have been created in some critical areas.

Hawaii boasts a few of the cases where species are being successfully reintroduced in the wild. *Hibiscadelphus giffardianus* was reduced to a single tree that died in 1930. Material from cultivated stock was reintroduced into the original habitat in the Volcanoes National Park. By 1968, ten mature trees and a number of seedlings had sprung up. Five species originally belonged to this genus but two, *H. bombycinus* and *H. wilderianus,* have become extinct.

The Lesson from Islands

Islands are closed systems, areas of limited space with burgeoning populations. They offer the continents a taste of the future. In fact, the future they portend is beginning to occur now. We can learn a great deal from the troubles they have experienced. Though islands are more sensitive to ecological variations and stress than continents, with enough effort even the continents can be totally degraded. Perhaps the most important lesson we can learn from islands is the need to have large enough preserves to counteract the biological law that spells out danger for species living on small islands.

Myosotidium hortensia

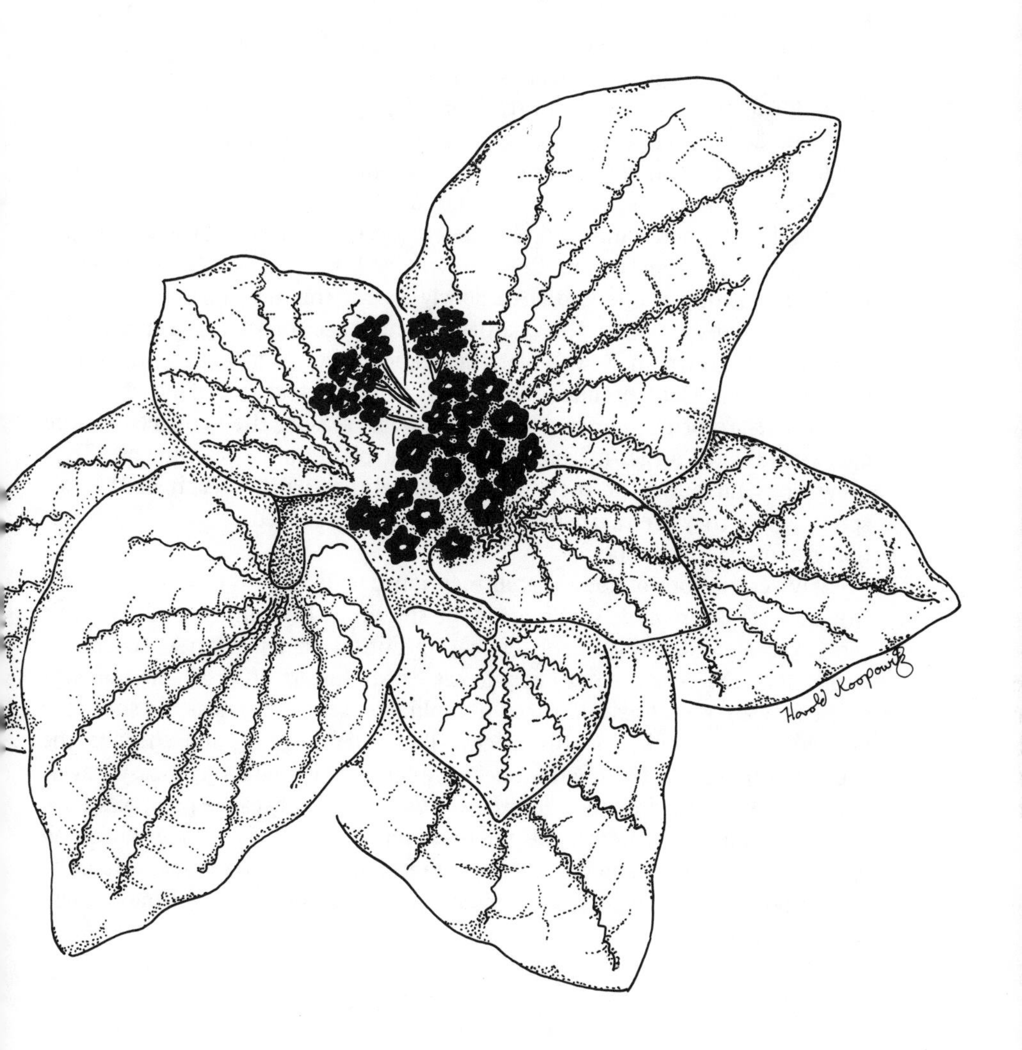

Case History: The Chatham Island Forget-Me-Not

Hidden in the temperate zone of the South Pacific, some seven hundred kilometers off the coast of New Zealand and due east of the city of Christchurch, are the Chatham Islands. Some six hundred souls eke out a living on this small cluster of islands, the largest being the 90,000-hectare Chatham.

This remote group of islands has seen at least three invasions. The first one (roughly 1,000 years ago) was by the Morioris, a passive East Polynesian people who adopted a nomadic existence as they hunted for food. The second invasion was led by the British in the late 1700s. This contact with the white man nearly decimated the Morioris. Less than one hundred years later New Zealand Maoris invaded the Chathams and enslaved the few surviving Morioris. The population of Morioris had dropped from 2,000 to 100 by 1862. The last Moriori died in 1933. The island's plants also fared poorly during the invasions.

Myosotidium hortensia, the Chatham Island forget-me-not, is a spectacular plant widely grown in New Zealand and known by gardeners in other mild climates. Each plant produces a rosette of quilted and ridged heart-shaped leaves that are bright glossy green in color. In season, the plant bears large clusters of blue forget-me-not flowers. The plants are ocean-loving. They grow on sand dunes or rocky ledges, sometimes within reach of the salt spray, where they are nourished by decomposing seaweed or the natural peat that occurs in those spots.

Early descriptions of these islands mentioned wide ribbons of land filled with true blue forget-me-nots that hugged the shoreline. Even in the middle of this century, people described acres of these beautiful plants. A person currently strolling by the ocean is not likely to see great clusters of forget-me-nots, for today they are rare. Pigs and sheep trampling and browsing near the shore are blamed for the decline of this species. Pigs relish digging and chewing the succulent tuberous stems, while sheep are tempted by the large, glossy leaves.

Rare or not, the prognosis for *Myositidium* is good because preserves have been set up on the smaller islands and the species is reported to be reestablishing itself. On Chatham Island itself, a private reserve has been developed. Here the largest population of plants is protected. We hope this giant forget-me-not will not be forgotten.

CHAPTER 9

Deserts

Deserts are a vital component of the earth's ecosystems even though these areas sometimes are mistakenly regarded as barren, relatively lifeless regions. Plant species native to desert regions and species that live in the semi-arid regions surrounding true deserts are both in trouble. The true desert plants are threatened when their habitats are destroyed by man's actions—whether from mismanagement, overgrazing or off-road vehicles. The plants that dwell in the adjoining, semi-arid regions are in trouble, too, since man's actions are expanding the boundaries of deserts and this expansion is pushing out the plants that thrive in semi-arid habitats, not arid ones.

The desert ecosystem looks rugged and robust, but these regions are literally brittle and fragile. The surface layers of soil in many deserts are not loose grains of sand but are knit together and cemented by lichen plants, soil fungi, and carbonate salts. The surface layer is not firm and can be broken easily. The soil is sandy and loose underneath. Once the surface mantle is disrupted, the soil below can easily be blown away. Therefore, winds can erode the desert; and rain, as sparse as it may be, can wash away the loose sand. Even in the "barren" strips of land stretched between obvious plants, the top few inches of soil contain a delicate webbing of slow-growing life. Some deserts have large trees and shrubs, but these may take decades or even centuries to reach their prime stature, a fact little appreciated by those driving through our desert preserves.

Desertification

About one-quarter of the world's land surface is considered to be desert. The natural deserts form two bands—one in the Northern Hemisphere and the other in the Southern—of lessened rainfall, in the 20 to 40 degree latitudes. Bounding the great deserts are semi-arid regions of savannah or steppe which, if not treated with respect, will themselves be converted to desert. The man-induced process, in which non-desert land is turned into actual desert, is called desertification. We have recently witnessed this process in the Sahel savannah, south of the Sahara Desert. The Sahara has crept southward at about nine kilometers per year. This might not seem very great to people who routinely drive long distances, but the desert will smother and absorb about 180 kilometers within twenty years. This can be expected to disrupt the economy of several central African countries and impose hardships on millions of people.

Desertification is not confined to Africa; the process is rampant in much of the Middle East and Asia as well. In this last half of the twentieth century, desertification has also occurred in the tropics. Despite tremendous amounts of rainfall, many areas of South America have been turned into deserts. Recently documented by plant scientists Drs. Fritz Went and V. Ramesh Babu, these new deserts include the area near the Rio Negro in the Upper Amazon, parts of the Lower Amazon where the rain forests were cut, and the Coatinga of northeastern Brazil. All of these regions receive at least 25 inches of rain per year and some areas are pelted with more than 100 inches each year. Once the forests were cut, the few nutrients left in the soil were leached out by the rain, leaving behind almost pure-white sand deserts.

Curiously, in the Far East, tropical forests have been replaced with highly productive rice paddies and other agricultural crops. In South America, Africa, and Australia, however, tropical forests are replaced with deserts. The reason for the difference lies in the history of the three old continents—all part of the original Gondwanaland, the southern continent. Through the millenia, most of the natural mineral nutrients were leached out of the continents. In contrast, the Asian areas are geologically younger and are comprised of rich volcanic soils that can support many crops.

In Australia, the sandy soil is so poor that, following a rain, much of the water soaks into the ground and then works its way to the sea via underground rivers. The desert conditions of central Australia are not so much related to low rainfall as to lack of nutrients. There was a stretch near Adelaide known as the "90-mile desert" about thirty years ago. Driving through the same area today, you see the rich farmlands. The Australians had found that the natural soils lacked copper, zinc and phosphorous. Copper and zinc are needed by plants in trace amounts. The Australians ploughed the desert, added the nutrients, and sowed the area with clover and grasses. These "improvements" coupled with the natural precipitation led to the disappearance of the desert. Good? Perhaps. With the desert went the natural, though impoverished, ecosystems of native plants and animals. So it seems that wild species can't win. They lose with either desertification or de-desertification.

One recent, significant case of desertification involves the Sahel, where savannah separates the dry Sahara Desert from what is left of the tropical rainforests. There are three bands of savannah; rainfall diminishes as you approach the actual desert. The Sahel receives only six to twenty inches of rain per year. The African countries of Mauretania, Mali, Niger, Sudan and Chad have considerable portions of their land covered by the savannah of the Sahel. The ecology of this system is sparse grassland interspersed with scrubby trees and occasional shrubs.

These savannahs are best suited to a nomadic way of life. There is not sufficient pasture to maintain prolonged grazing; a nomadic system permits the range to recover after it has been grazed. A number of events—both foreseeable and preventable—caused the ecological catastrophe of the Sahel. To begin with, the human population rose from 18 to 25 million. This increased the demand for food. To solve this problem the cattle population was increased from 15 to 25 million. Fritz Went and V. Ramesh Babu, botanists, report that earlier native peoples had kept the cattle population down to 15 million, but that "computer intelligence" determined that the area could stand 25 million head of cattle. The locals were encouraged to increase their stock. This increase occurred during a run of good rainy years, but the subsequent several years of severe drought were not totally unexpected. The needs and pressures of extra people and extra cattle turned much of the Sahel into the shambles it is today. Roughly 100,000 people died from starvation during the years of drought, 1968-73, and people there continue to suffer today. The extent of desertification was horrifying. If the Sahel had been properly managed, it could still produce beef for much of Central Africa.

Unless supplemented with water and fertilizers, savannahs, by their nature, are fit only for nomadic people. These areas always will be riding on the brink of disaster, since prolonged natural droughts do occur. Where savannahs have been fenced and protected, the sparse vegetation returns and can be carefully grazed. They must be managed carefully. Haphazard overgrazing just leads to further desertification.

Desert Dwellers

Plants have ingenious ways of dealing with situations in the desert. Cacti and other succulents carefully collect and horde stores of water and protect the moisture from transpiration. These plants develop two kinds of root systems: one system has deeply penetrating tap roots that delve down in the soil looking for moisture; the other system has a shallow network of roots that opportunistically waits for surface waters.

Many arid areas of the world do not get sufficient precipitation, but they are exposed to fog banks. This is particularly true of coastal deserts situated on the west coasts of continents. Plants in these areas frequently have very fine, needle-like leaves that can effectively intercept fog droplets. These droplets run down the leaves and then drip onto the soil around the plants, where the surface roots are able to suck up the water. Even without much rain, enough water can be absorbed from the air to support forests and vegetation. The coastal redwoods of California certainly do not grow in deserts, but they are dependent on fog drips for much of their water. In South America, however, patches of forest are able to survive in rainless areas. Lomas, which are vegetated areas surrounded by deserts, occur in South America along the west coast. These are fragile areas where trees and shrubs take a long time to grow because the only source of water is fog. Many lomas now are being desertified.

The most remarkable desert forest occurs in the Pampa del Tamarugal in a Chilean desert where the rainfall is less than 1/20th inch per year and there is no appreciable underground water. Despite this, the Chilean Mesquite (*Prosopis tamarugo*) forms trees with trunks three feet in diameter and more than sixty feet high. How do they manage to grow so well on such miniscule amounts of water? Apparently leaves absorb moisture from the air during the night and then transport the water through the plants and then down and out the roots. Moisture is available for transpiration during the desert day. Besides this night-time procedure, any fog that drips to the ground also helps the trees' growth. Nature has thus fashioned the means to sustain a tree in a most intricate way.

California's Deserts

While deserts in Third World countries are being demolished by overgrazing and resource mismanagement, deserts in more affluent countries face more exotic forms of destruction. Off-road vehicles (ORVs)—essentially recreational four-wheel drive cars, dune buggies and dirt bikes (motorcycles)—are used by Californians and others in the southwest United States in ever-increasing numbers. In the process of providing recreation, these vehicles destroy much of the life within the fragile deserts. More than one hundred species in California's deserts have found their way onto the state's list of endangered and rare plants.

The Californian deserts fall into three main groupings: the high deserts, the Mojave, and the Sonora. The high deserts range in altitude from 4,000 to 8,000 feet and extend to the Great Basin of Utah, including the Nevada desert. Only a corner of high desert makes it into California.

The Mojave desert, ranging between 2,000 and 4,000 feet, consists of closed basins surrounded by high mountains and makes up most of the desert area of California. The flora in the Mojave is quite rich and can be spectacular after the sparse winter rains, which seldom consist of more than two or three inches. Some 757 known species—including 51 endemic species—live in the Mojave.

The most prevalent species is the scrubby Cresote Bush (*Larrea tridentata*), but many regions also display the Joshua Tree (*Yucca brevifolia*) that sets the character of the region. Many of the species are ephemeral—short-lived annuals that grow rapidly and mature after the winter rains, often producing their seed before spring is underway. These ephemerals are the flowers that Californians drive to the desert to see at the end of winter.

The Sonoran Deserts are low elevation regions, ranging from Death and Panamint Valleys to Anza Borego, then across to Arizona and into Mexico. Like the Mojave desert, the dominant vegetation is Cresote Bush but trees are present, especially along the dry water courses. Palms are found in oases. Saguaro cacti used to extend into California's deserts but, by now, all have been destroyed. Panamint and Death Valleys have 454 flowering plant species, of which 199 are endemic.

Destruction Derby

More than half a million four-wheel drive vehicles, as well as more than 200,000 dune buggies, are owned and operated in California. In Southern California alone there are more than one million motorcycles—each one capable of tearing up wide swaths of vegetation.

On most weekends 5,000 dune buggies can be found on the Algodones Dunes. In one recent year more than 151 competitive ORV events were held involving 67,000 participants and 189,000 spectators. The dust raised by these events is so thick that sometimes it can be photographed by orbiting satellites.

One race from Barstow to Las Vegas, called the Hounds and Hare Race, drew more than 3,000 contenders. The 150-mile race is held in two heats with a starting line that stretches one mile wide. Its effect on the desert can only be described as rampant destruction. The destruction was counted and assessed after one Hounds and Hare Race. The damage? 140,000 cresote bushes, 64,000 burro weeds (*Franserria dumosa*) and 15,000 yuccas demolished within a 100-mile stretch. These figures do not include scarce or even rarer plants that were destroyed. As early as 1968, desert areas with nearly 60 percent of their perennial cover destroyed were apparent. Because they are desert plants, the shrubs take decades to regenerate. All the small animals and reptiles that use these plants for shelter, as well as the ephemeral herbs associated with the shrubs, are lost. As the ORVs race over the ground, the soil mantle is broken or compressed. The micro-organisms and insect larvae hidden in the soil are lost. Most ORV events are held in the spring—prime reproductive time for both animals and plants—when the shrubs are achieving their skimpy growth, the ephemerals are producing their seed and the animals are building nests. The destruction could not happen at a worse time.

In 1973, the U.S. Bureau of Land Management came up with a plan for managing the deserts. About 18,000 square miles were allocated for unrestricted use by ORVs, with just 1,069 miles closed for their use. Because the ORV users complained, 24 percent of the closed area was re-opened to them. Well-organized, well-monied, and vocal ORV users claim that they are the only ones who use the deserts. Their contention is false but, unfortunately, good estimates of the other users are not available. These others include garden clubs that visit the wild flower displays, schools, photographers, rock hounds, campers, and others. These other users are not organized and are not connected to entrepreneurs such as the ones who make substantial profits from organizing ORV events.

Of all the Californian deserts, the most fragile are the sand dune systems. Of the five major dune systems in the desert, ORVs have access to four, and the fifth is only nominally closed. Another dune system along the San Joaquin River, the Antioch Dunes, has now been declared a preserve for endangered plants and insects. Two of the plants that occur—*Oenothera deltoides* var. *howelli,* a desert primrose, and the Contra Costa wallflower, *Erysimum capitatum* var. *angustatum*—are endangered species under both federal and Californian law and were depicted on postage stamps in 1979. The primrose has taken on new significance since evening primroses now appear to be the source of a potent new drug. The Antioch Dunes, while some distance from the nearest true desert dunes, share similarities in the fauna. In 1908, a U.S. geological survey estimated 190 acres for the dunes. Aerial photography in 1952 showed that the eastern two-fifths had been displaced by heavy industry and that the next third was being mined for sand. When federal ownership finally was assumed, only 3.7 acres had anything resembling the original ecology. Among the first problems to be faced involved the enforcement of restrictions on ORV use. We now wait to see if it is possible to return the Antioch Dunes back to its previous condition. To do so will require active management and replanting. In many ways this pioneering project should be closely watched by the rest of the nation. If successful, it could be carefully repeated elsewhere.

We believe that desert plant species represent an untapped treasure chest ranging from industrial chemicals to oils and fuels to new medicines. The potential is vast, but a few dangers do exist in the exploitation of these plants. Because most desert plants grow slowly, when hurried along with extra water and nutrients they may not produce their treasured chemicals in the same proportions. One attraction of desert plants is the expectation that they can be grown in arid and semi-arid regions. However, as more acres are brought into cultivation, the stresses on wild desert species diversity will increase. Of the federally recognized endangered species in the United States, about half are plants adapted to desert conditions.

Obregonia denegrii

Case History: Obregonia

As the dry, hot wind whistles through the yuccas and jatrophas, it tries to wrest precious moisture from the desert scrub. Some plants, including a special cactus called *Obregonia denigrii*, protect their resources by burying the water underground. *Obregonia denigrii* snuggles into the stony soil, hiding most of its body from the hot, dry air. It hordes its drops of water in an underground stem.

Obregonia is rather uncactus-like. On top of its thick, fleshy subterranean stem is perched a rosette of grayish pyramid-like tubercles. Each tubercle is actually a wedge of tissue bearing a tuft of soft spines. At the center of the rosette, developing tubercles are protected by white wool. In the correct season, white flowers emerge from this region of the plant. The cactus is scarcely four inches in diameter, although very old specimens can be larger. Because the cactus has such an unusual shape, it is considered a connoisseur's plant—one sought after by those who specialize in desert plants. The shape seems to have predestined the species to suffer an untimely end.

O. denegrii, the sole species in the genus, occurs in a very restricted area in the state of Tamaulipas, Mexico. These plants can be found on the stony mountainside of the Sierra Madre. Because the species is avidly sought by hobbyists, there are ready markets for them in Japan and Europe. Individual plants are valuable but they do not command excessive prices. Nevertheless, they can be a source of extra income for the local Mexican villagers.

Unfortunately, collected plants do not always end up in cultivation. For a variety of reasons, wild-collected specimens may not get sold and are left to die after being uprooted. One eyewitness report described roughly one hundred heaps of *Obregonia* plants. Each pile contained between thirty and forty plants, all in various states of decomposition. The plants had been uprooted from their wild nesting places in the mountains, but for unknown reasons had never been sold. Nearly four thousand plants, a sizeable percentage of the wild population, was needlessly destroyed. This is a regretable occurrence for any species, but for a cactus as rare as *Obregonia*, it is a tragedy that hastens its end. The Mexican government has barred exportation of its indigenous cacti except under license, but this does not mean that collection is illegal. Furthermore, cacti can be smuggled out by tourists, poachers, or wild species jobbers.

What is the future for this species? Its future in the wild is doubtful. There is no trick to growing *Obregonia* from seed; it grows quite easily. However, like many of the choicer, slow-growing cacti, it has been simpler and cheaper to pull plants out of the ground in the wild than to grow them from seed.

CHAPTER 10

The Temperate Zones

The mild nature of temperate regions sometimes makes these areas seem immune to serious trouble. Nevertheless, temperate countries suffer the burden of industrialization and its offspring, pollution. These regions also bear extensive agricultural and burgeoning population pressures. Some people in Western countries tend to feel smug in their corner of the world as they learn of the destruction in the tropics. The truth is that nearly all Western temperate countries have devastated their own ecological systems. The forests that once clothed Europe were cut down for fuel—even before the industrial revolution occurred. The plateaus of Spain and the hillsides of Greece all once were covered with trees. These areas are now mere rubble and sparse grass. Even among the rocky sectors, the remaining plants are severely threatened. The great prairies are gone in the United States, mostly due to the farms that degraded the land. The large game animals that roamed both Europe and North America have already disappeared or now stand in jeopardy.

No wonder both the developed European countries and the temperate colonies they established now hunger for the wild. The forests and the prairies and the enormous animals are all part of a heritage already lost. The rest of the world is destroying what is left.

The European Situation

In Europe, some areas have been hit particularly hard, with nearly half of the endemics of those regions currently threatened. Lists of threatened plants in Europe provide estimates that nearly one in five species is jeopardized to some degree. Temperate countries in both the Northern and Southern Hemispheres are among the richest and most enlightened nations in the world. We expect people in these areas to understand conservation problems and play leading roles in seeking solutions. As expected, we do see a rising consciousness in these countries. People in these areas know their floras best and were the first to begin publishing regional Red Data books which list and describe endangered or threatened plants in those regions. These books have helped trigger substantive efforts (such as gene banks) aimed at reducing the problem.

A French botanic garden, the Conservatoire Botanique de Porguerolles, has recently been established to conserve French plant species. The primary task of this garden is to collect seed of threatened plants, grow the plants, and then produce enough seed to restock wild areas. The French also are concerned with maintaining genetic stocks of cultivated crop trees such as mulberries, figs, and olives. The Mediterranean area used to be exceptionally rich in plant species but, similar to other Mediterranean-like ecosystems in the world, plants in this area are among the most threatened. The French Riviera attracts just too many tourists, industries, and builders for plants to be safe.

Most countries in northern Europe now are concerned with saving their endangered floras. Poland has only six endemic species of plants with another five species occurring in limited areas outside Poland. Fortunately, the Polish Academy of Sciences has now set up a gene bank for all endangered species in Poland. The country has 614 nature preserves, of which 370 are intended primarily to save plant communities. Those involved with the Polish reserves have learned that active management of the reserves is important or trees will crowd out meadows. When grazing animals were removed from the Tatra National Park, the mountain meadows were filled in by encroaching trees and shrubs which threatened the smaller meadow flowers. Endangered species in Poland are protected by law. However, the plants listed include only about half the number of endangered species. Officially, in 1978, 141 species were legally protected of the more than 2,000 plant species that occur in Poland.

The Soviet Union

The USSR covers some of the largest land mass on the planet. In that country, which ranges from tundra to desert and grassland to mountain forest, about 21,000 plant species grow. Of these, two hundred species face imminent extinction and about two thousand others have populations sufficiently reduced to warrant listing as threatened. More than 115 botanic gardens in the USSR cultivate about one-third of Russia's endangered species. As concern for threatened plants appears to be growing, many of the small republics have published Red Data books listing threatened plants in specific regions. It is not surprising that the largest number of officially protected species—about four hundred—is found around Moscow.

As in other parts of the world, land in the Soviet Union is ploughed to help feed the burgeoning population. One botanically rich region is Kazakhstan, where there are 5,630 known plant species. For some unknown reason only 286 species are listed officially as endangered, a surprisingly low percentage. Compared to other regions of the world, we would have expected at least double that number. The Kazakh area has given us many bulbous plants, such as *Tulipa greigii* and *Tulipa kaufmanniana.* These two species were significant in the development of modern tulip varieties. One of the most interesting plants, *Ostrowskia magnifica,* also comes from this area. This plant produces stems up to eight feet tall, bearing lavender bells six inches in diameter.

The Georgian Republic contains about 1,300 species of plants and has contributed widely to other gardens—providing delphiniums, paeonies and scabiosa, among others. About four hundred Georgian species are considered endangered, which falls in line with expectations.

There are some local controls in the various republics of the USSR, but the use of natural resources is centrally planned. The Russians have a long history of adding and subtracting from their system of reserves. Over the last half-century or more, the Soviets have developed a network of natural areas called *zapovednikii.* The word literally means "restricted area." These areas range from preserves for scientific study to expanses that resemble the American national parks. There are areas of wilderness besides hunting and fishing preserves. The laws authorizing *zapovednikii* as specifically being "forever withdrawn from economic utilization for scientific research and cultural-education purposes" underscore scientific importance. Most *zapovednikii* have their own research labs, staff, museums, and publish their own journals.

Unfortunately, "forever" has different meanings. In 1951, there were 128 *zapovednikii.* Somehow there was a drastic decline and revision in 1952. Suddenly there were only forty *zapovednikii.* The number built up to 93 in 1961, but by 1964 it had declined again to 66. The number has increased somewhat since then. Sometimes the same area is classified and re-classified a number of times. We won't know until later the exact effect these frequent changes have had on these regions.

Although Lenin gave official sanction to *zapovednikii* in 1921, the system was begun much earlier. By the time of the revolution, seven preserves had been set up and an additional two were created between 1919 and 1920, just after the revolution. The Central Forestry Department of the Agricultural Commissariat classified large areas of forest off limits to logging in 1921, calling this land "territories of future national parks and monuments of nature. . ."

The main shortcoming of the Soviet system is the ease with which *zapovednikii* can be abolished. One simple example shows the dangers. The Kronotsky preserve in Kamchatka resembles Yellowstone in that it is one of the world's greatest geothermal resources, containing many large geysers. The preserve was abolished in 1951 to permit prospecting for oil. When none was found, the preserve was re-established—but not for long. The preserve status was abolished again in 1961 to allow for the construction of a power station. Then, preserve status was re-established again. Not even our Secretary of the Interior would attempt to do that to Yellowstone National Park.

The U.S.A. Situation

The United States, world leader in conservation, was one of the first countries to set up a system of national parks and national forests. Before the national effort, scattered individual states had attempted to set up their own conservation laws. One of the earliest states to do this was Pennsylvania, where in 1681, a law required one acre of trees to be left behind for every five acres cut. This law of the early settlers is only now considered wise legislation. At the federal level, Yosemite and Mariposa Big Trees (*Sequoiadendron sempervirens*) were protected by law in the 1860s, but it wasn't until 1872 that Yosemite was declared the nation's first national park. This kicked off the grand movement to save spectacular scenic wonders for the general public. The process of setting up parks and wildlands usually requires a battle, since entrepreneurs are always waiting to exploit those regions. Even after legislation occurs, new laws can reverse previous gains. Many people still see progress as a taming process, where nature is subdued by the will of man. This is the same mentality that assumes that everything on the planet has a purpose which is solely to serve man. Such people fail to see man as a cog in the ecosystem—a cog that both affects and is affected by the environment. The protection laws have brought some relief, but the results have been spotty. While we were busy protecting the redwood trees in California, we killed off nearly six million buffalo on the prairies.

The use and abuse of national parks has changed over the decades. Until recently, national parks were considered recreational centers. The public usually was given access; if the parks were closed, it was only because of hazardous conditions. Parks in those years were primarily for people. Today, we know that parks must be managed to be effective and that this includes managing the visitors to the parks. Private cars are banned in some parks. In others, the numbers of people entering or obtaining campsites is limited and reservations need to be made well in advance. Elsewhere, visitors are bussed through while the tour conductor explains the natural history of the park. This is all part of the plan to limit access. While this is not always convenient, we will have to put up with these limitations if parks are to withstand overuse and abuse. The idea that parks can be recreational areas for human use, or are places where urban-dwellers can renew their bonds with nature, is limited by the burden our numbers place on the fragile ecology of the parks.

National Forests

It was the forests and woodlands of the United States that first impressed the early settlers, who were greeted with an ocean of leaves and branches that stretched indefinitely to the west. There are estimates that in 1630 about 950 million acres of forested land existed in the continental United States. More than two-thirds of these forests were in the east. Before long, however, the forests ran into trouble with the settlers since wood was used to construct houses and keep them warm while the forested land was needed to grow crops. Slowly the woods were converted to farms. Sometimes forests claimed cleared land, but usually it was the other way around. By 1920 about 350 million acres of forest had been cleared. Wood continued to be a major source of fuel until about 1933, when electricity became widespread. Since the use of forest wood for fuel has declined, the trees have had a chance to recover, a partial reason why forested land has experienced some regeneration. More than 130 million acres of non-commercial forest have been added to the nation's inventory since 1930.

The national forests system tends to be confused with the national parks system in the United States. In reality, there is a world of difference between the two. The parks systems is primarily for recreation and conservation. The national forest system is a reserve where resources are available for exploitation. Lumbering occurs in the forests along with exploration for oil and gas and strip mining. Recreational activities—camping, hunting, fishing—can take place, too.

Theoretically, the forest is a renewable resource. Nevertheless, there is a curious philosophy about the national forests that actually runs counter to the idea that forests are renewable. The belief is that if mature trees are not cut down, they will be lost to decay and insects. This belief is true, but resources are renewable only through recycled matter. The old decaying trees and the flesh of the insects are recycled into new growing trees. You cannot harvest trees indefinitely from an area without permitting recycling to occur. It is like trying to grow many harvests of corn without fertilizer. You can't get something from nothing—not even national forests.

The forest service allows clear cutting—the removal of every tree in a specific area. If the trees are replaced, this is generally done with a monoculture. Conservation groups managed to stop clear cutting in the courts in 1973, but it was not long before congress granted the National Forest Service the ability to again allow this total removal. Big money interests are powerful. The recent housing industry slump in the United States has reduced much of the lumbering pressure in the national forests for the time being; however, this is just a temporary situation.

Prairie Grasslands

When the first settlers emerged from the New England forests, they found a vast grassland that stretched toward and beyond the horizon. It was a land with seasons: harsh, snow-packed winters; short, brilliant springs; long, bright summers; and fall, when the grasses and perennials dried up and left behind tinder for prairie fires. The prairies consisted of a surprisingly limited number of species—just two hundred or so, of which about eighteen were grasses. Most of the rest were perennial plants, whose flowers sparkled against the prairie grasses during the growing seasons. Trees and shrubs clustered at the edges of rivers.

There is no agreement on whether the prairies were a natural state or a forced ecology caused by the annual fires (perhaps of Indian origin), or the hordes of buffalo that migrated freely, grazing where they wished. In any event, this ecosystem existed for thousands of years. Within 150 years, Western man has destroyed the system. Some say the prairies were tamed by the plow, others insist that they were destroyed by man's ecological incompetence. The reduction of the prairies is dramatized in Illinois. The state originally contained 40,000 square miles of prairie. A 1977 survey revealed that this land had been shrunk to a mere six square miles.

Although Iowa purchased Hayden Prairie in 1945 and now has incorporated it into the state preserves system, it was not until the mid-1970s that anyone seemed to care whether or not prairies were saved. A Tallgrass Prairie National Park has been designated, too. Meanwhile, other smaller prairies are being set up in arboreta and various institutions. One of the most interesting is the prairie system within the accelerator ring at the Fermi National Laboratory nuclear research facility in Batavia, Illinois. This prairie encompasses seven hundred acres which have been replanted with many natural prairie species. It was the recognition that prairies thrive only when they are burned that led to their successful reconstruction and restorations. What a shame that the remnants left will be mere shadows of the great grasslands that covered the midwestern states.

A curious sidelight to the quest to reconstruct prairies is the unusual repository of prairie flowers found in cemeteries. When the early settlers spread into the prairies, many died from the hazards of prairie life and natural causes. They were buried in small cemeteries where among the crumbling gravestones, many prairies species still survive. Conservationists now comb through these old cemeteries looking for the original grasses and perennials that used to grow across the great grasslands.

California

California, one of the richest states in the nation, claims leadership in many fields, not the least of which is the number of plant extinctions. More than ten percent of the endangered plant species in the continental United States are from California. Along with its Mediterranean-like climate, California also has many different geographies. There are the deserts and the high Sierras, the inland valleys and coastal hills, the forests and grassy plains. This geographical variation in turn leads to many, many kinds of habitats and many different plants adapted to each. Some of these plants are naturally rare and have never been found except in small isolated populations. Unfortunately, since the metropolis of Southern California stretches from Santa Barbara through Los Angeles and down to San Diego, many of these small habitats and their plant species disappear as the cities gobble up land. Often, endemic species are destroyed even before they are recognized. Air pollution has a devastating effect on the vegetation as well. In the San Bernardino Mountains east of Los Angeles, the national pine forests are dying out, victims of the noxious air. The trees die faster than they can be replanted with smog-resistant species. Unfortunately, by their very nature, smog-resistant species are different from the natural ones. We consequently must expect the ecology of the forests to change. The effects of this change on other normally occurring species cannot be predicted.

California has a delightful flora with many showy plants of tremendous garden potential. There are active groups and foundations trying to promote the use and conservation of native plants. One of the marvelous vegetation systems in California is the open oak woodlands where scattered oak trees stand in open fields of grass. Small bulbous and herbacious plants scatter their flowers across these fields in the late winter and early spring. This system is adapted to a natural fire ecology. Man, however, finds fire to be a danger and for many years has strenuously "protected" the plants from natural blazes. Fires in open grasslands are fast and do not burn very hot. They will not harm the mature trees but they do kill most of the oak seedlings. This is what the natural ecology requires. Protection against fires has allowed so many oaks to germinate that the grass has given way to oak scrub land. Now, when a fire rages through the area, it is hot enough to kill the mature trees as well as some of the underground roots, tubers and bulbs. Only in the last few years has man realized that controlled burns are necessary, not only in the open grasslands, but also in the characteristic *schlerophyl* vegetation of the coastal area.

Swamps and Soil Erosion

Traditionally, one of the great signposts of progress has been the drainage system used to dry out swamps. These drainings have been major projects for centuries. Draining a swamp was considered a way to improve the land—turning it from a useless, disease-ridden area into lush farmland or land extensions upon which cities could expand.

The swamplands and wetlands of Europe have nearly all been drained and so has much of the American midwest. We now know that swamps are far from useless. They are the most productive ecosystem in terms of photosynthesis—even greater than the tropical rain forests. In Indiana, the original Kankakee Swamp was described in literature as being a "truly remarkable wildlife paradise." This swamp was drained and turned into agricultural land. For two years the land bore good harvests, but at the end of that period the water table had dropped so low that the land was worthless for farming. The area's ecology had been destroyed—its rich biological diversity replaced with a few weedy, grassy species.

Drainage is not the only villain to plague wetlands and marshes. A new enemy is now on the loose—a plant that can suffocate even cattails, rushes and sedges. *Lythrum,* also called purple loosestrife, made its first appearance in a New York preserve in 1951. Five years later, the species covered a single acre of dense vegetation. Purple loosestrife now covers thousands of acres, squeezing out other wildlife, and has spread into Canada and as far west as Minnesota.

There are more subtle ways that the land has been changed. Topsoil is a resource that renews very slowly. It takes an enormously long time to make an inch of good earth. Between 1850 and 1950, the United States lost more than 50 percent of its topsoil. In some of the southeastern states, more than 75 percent of the topsoil has been lost. Soil is usually washed away, ending up as silt in rivers, but during extensive droughts it also can be blown away. Soil is still being lost at the rate of one billion tons each year. This loss eats at the heart of the United States' greatest strength—its agriculture.

The Southern Picture

Not only has the North American prairie disappeared, but its South American equivalent has met the same fate. The Pampas of Argentina has been almost totally destroyed by cattle farming and extensive agriculture. Biologists from southern South America have pointed out the radically different landscape as compared to what the first settlers found.

In South America we see that the creation of preserves does not necessarily protect an ecosystem. Argentina has between fourteen and eighteen national parks or nature preserves. Until 1976, most were considered recreational areas, not preserves to protect the biological community. Outside of Buenos Aires is the Punta Lara forest which was set up as a biological preserve about forty years ago. *Equisetum giganteum,* a primitive sporebearing plant which was once common in the forest, has continued to be plundered and now is quite rare. Without real protection, the Punta Lara forest has lost trees to lumbering, and smaller plants have been dug up by weekend picnickers. Two foreign intruders, *Ligustrum lucidum,* a privit, and *Rubus ulmifolius,* a blackberry, are replacing the natural vegetation of the forest. These aggressive invaders need to be controlled if the forest is to be saved.

In the Patagonian Andes, Argentina has several national parks where the forests are dominated by *Nothofagus,* the Southern Beech. Hunting was permitted in the parks. When the native deer were killed off, the Argentinians introduced European deer. Despite the hunting, the European deer did very well. Unfortunately these animals love to eat beech seedlings and young trees, and strip the bark of older trees. The forests now are in jeopardy.

In former years, there were deciduous forests at the base of the Anconquia Mountains, but these have long since disappeared. Some trees can still be found along streets and in parks that used to be part of the forests. The names of some of the trees, *Tabebuia* and *Tipuana,* will be familiar to northern gardeners. Whether or not these trees continue to be propagated and planted in the towns of those areas is not known.

Schinopsis, the dominant tree of Argentina's Chaco woodlands, is nearly extinct. These trees were used for construction and to fuel the railways. The regeneration of the Chaco has been thwarted by goats that range freely over the land. Some of the Chaco was turned into vegetable farms, but the drier regions reverted to desert. Other plant problems exist in Argentina and there is a similar troubled situation in Chile. We hope that these countries soon will begin to respond to the ecological distress that has been generated. Perhaps the sort of awareness now emerging in Brazil and other tropical American countries will spread to the southern temperate regions.

Two other major temperate climates located in the Southern Hemisphere are floristically important: South Africa and Australia. Unlike South America, both of these countries have intense pride and awareness of their native flora. But this awareness does not guarantee protection. The flora of South Africa, particularly the Cape region, is exceptionally rich. The number of plant species at the Cape varies according to who is asked. In conservative terms, six thousand species have been counted, but claims for more than ten thousand species have also been heard. In the most recent 1982 update, 1,621 species were listed as either extinct, threatened or critically rare. The richness of the flora and its jeopardy follows the same boundaries as the metropolis of Cape Town, its suburbs, and surrounding farms. This area is less than one percent of South Africa; yet it contains 65 percent of the country's endangered species.

Some famous plant families are particularly threatened here: one-third of the protea family and more than one-fifth of the iris family are endangered. Besides encroaching civilization, aggressively invasive Australian weeds are literally crowding out the native species. The South Africans have strict legislation to protect the flora. A series of botanic gardens and preserves dedicated to the natural plants of the country has begun. The national pride in the native plants is also important. Many plant societies are devoted to wildflowers and plant nurseries tend to specialize in indigenous plants. We can expect a fair number of species to be saved due to this nationwide interest, but the problem is still very real and won't get better. One authority considers one-quarter of the Cape flora to be endangered; just 25 percent of these are in cultivation in the major botanic gardens of the world, including the famous gardens in South Africa.

While the South Africa flora has produced many of the great garden plants of the world, the equally colorful Australian flora seems to be far less well-known. Australian lists enumerate some 2,206 threatened plant species—about 10 percent of the continent's flora. The percentage of endangered species in Australia is lower than in many other countries, partly due to the concentration of populations in relatively few large cities. However, areas where the number of endemic species is high—Western Australia and Queensland—appear to be problem areas, with human activity threatening the flora. Among the plants most threatened in Australia are species of *Eucalyptus* and *Acacia*. This is rather ironic since other monoculture species of *Eucalyptus* are replacing the tropical rain forests and *Acacia* species are among the noxious weeds threatening wild South African species.

Dianthus gratianopolitanus

Case History: The Cheddar Pink

In Somerset County, England, there is a deep gorge flanked by immense limestone cliffs. There, perched some fifty feet above the floor, the Cheddar Pink makes its home. The plant grows dense cushions of needle-like leaves that may be gray-blue or silvery-green. The leaves carefully hug the cliffs and from this base six-inch stems rise up bearing masses of inch-long, jagged-edged pink flowers. The fragrance of the deliciously scented plants fills the air around them. These days, the Cheddar Pink clings to inaccessible precipices among the broken rock, fighting off extinction as much as it is fighting off gravity. In former years the cliffs were coated with a vast foam of plants. During this century, the accessible plants have all been torn off to satisfy the needs of the world's rock gardeners. For the last fifty years, the Cheddar Pink has hovered at the brink of extinction. The plant is actually not alone in its troubles. England has lost 10 percent of its native plants over the last century and perhaps another three hundred species are threatened.

The Cheddar Pink, which draws its nickname both from its home territory in England and the pinked or jagged edges of its petals, has the misfortune to be properly named *Dianthus gratianopolitanus.* Most alpine and rock garden enthusiasts shun its tongue-twisting name and instead refer to it as *D. caesius.* This species, a relative of the carnation, has contributed to the modern garden pinks. In its own right, the Cheddar Pink is ranked near the top of the list desired by rock garden enthusiasts. The plant has been nearly decimated in the wild, but plant collectors need not have disturbed the species in its natural habitat since it is easy to grow and can be readily obtained from seed. This flower will even survive in the cracks of a garden wall. We can only hope that the remaining rose-pink summer foam of flowers will continue in peace and that perhaps the plants will recolonize the Somerset Cliffs.

D. gratianopolitanus is not confined to England, but also occurs in France and Switzerland. Even in those locales the Cheddar Pink has been over-collected and is now rare. Other species of the *Dianthus* genus are also in trouble. Some fifteen species of *Dianthus* are listed among the rare and threatened European species in cultivation. Many other species may not even survive to find refuge in botonic gardens.

CHAPTER 11

Green Cancers

Despite the enormous threats mankind and even wild animals present to plant species, plants sometimes are plagued by other plants. In most cases, troublesome plants are considered weeds. The definition of a weed is not precise, but generally, plants referred to as weeds are ones for which mankind has not found a use, or which grow where they are not wanted. Weeds are ubiquitous plants that seem able to survive and spread despite man's efforts. Many of these unwanted plants are superweeds—plants that grow so rampantly that they resemble science fiction fantasies. Almost before they are detected, superweeds smother the natural species and quickly cover acre after acre with useless vegetation.

Sometimes the competition from superweeds is subtle, with one plant growing here and another one there. The total soil surface covered by these plants soon adds up. Other weeds, such as Kudzu, are vines that cover a vast number of acres, literally smothering the natural vegetation. The vines clamber over the native trees and shrubs, preventing sufficient sunlight from reaching the photosynthetic organs of the host species. Other plants grow so closely together that native species are squeezed out.

Sometimes superweeds are plants growing out of context, situated in locations without natural enemies. Frequently, this happens when these plants are moved from one continent to another. Besides lacking enemies, these plants adjust to a variety of growing conditions, such as extreme conditions of temperature, soil and water.

Millions upon millions of dollars are spent each year to control superweeds. Because herbicides are only partially successful, superweeds are often eradicated with biological controls. Such measures usually are thwarted in the end because most of these superweeds produce vast quantities of seed that will germinate years later and reinvade the areas. Another method of control is the use of insects, which will eat growing shoots and seeds. This can be the ideal method, provided care is taken to keep the insects from attacking the native species or agricultural crops. Weed problems usually are greatest in the tropics. In temperate regions, weed controls can cost 10 to 15 percent of the crop's value. In the tropics, however, roughly half of the cost of the farming effort must go toward weed control. Farming is impossible in some tropical areas due to superweeds. The search for more effective weed controls must be expanded.

Weeds get started and spread in a variety of ways, although often their beginnings can be traced to some activity of man. Some out-of-control species are actually manmade hybrids, but many are just naturally aggressive species. Before the days of aviation, birds carried seeds from one country to another in the mud on their feet. The wheels of airplanes and jets are modern carriers.

Hybrids Gone Crazy

Genetic engineering technology is about to explode upon our modern world. The initial fears that scientific experiments could go astray and release dangerous manmade organisms have quieted down. We have not had to contend with the creation or escape of any frightening manmade creatures; so genetic engineers are once again permitted to pursue their experiments. While the earlier fears appear to have been overrated, we can look back and see how early "genetic engineering" with plants led to *Lantana camara,* one of the most noxious weeds of the world. This weed is rampant in the southern hemisphere—occurring throughout much of South America and ending up into the southern areas of the United States. The plants cover acre after acre in Africa, south of the Sahara, and are common in Madagascar, India, Southeast Asia and Indochina. From there, *Lantana* extends south through the Indonesian archipelago and into tropical east Australia, where it has moved across the islands of the Pacific and into Hawaii. *Lantana* crowds out native species, invading both pastures and inaccessible areas. The plants are toxic to livestock but produce berries which birds eat and spread. The distribution of the species appears to be limited only by its inability to tolerate winter frosts and snow.

Lantana was created using a complex of West Indian and South American species taken to Europe at the end of the 1600s. The species were hybridized (mainly in France) to make attractive summer bedding plants. More than six hundred different hybrids were made. The colorful flowers are like verbenas and made delightful mounds of showy spikes in the summer. Being tender, *Lantana* were carefully grown and protected in greenhouses during winter. The extensive hybridizing mixed and altered the genetic makeup of these plants, which probably accounts for the weed's great adaptability and variability.

Enterprising gardeners soon realized that since *Lantana* hybrids provided summer color in temperate areas, they would be fantastic garden plants in the tropics. Before long, manmade *Lantanas* that would love hot, humid climates were headed back to the tropics. By 1860, colorful *Lantanas* were displayed in private and public gardens from Calcutta to Cape Town, from the Gold Coast to Tahiti. The plants reveled in the warm sunshine, producing colorful layers of flowers as they clambered and grew luxuriantly. Butterflies pollinated the plants and soon blackberry fruits glistened among the bright colorful blossoms. Birds who came to feast on the berries flew off to spread the seeds.

Man unwittingly had created, and then let loose upon the world, a green plague. While the early *Lantanas* were beautiful, after a few generations in the wild, the brighter colors were lost. The newer flowers were much less exciting, but they were wildly successful. They seemed able to change their gross shapes to adjust to the environment and thrived just about everywhere.

The search for insects to combat *Lantana* has been extensive, but nothing definitive has been found. For a while, it appeared a fly fed on the seeds, but it turned out that the flies left the seeds intact, preferring instead the juicy part of the fruit. Other insects prefer different groups of *Lantanas*. Two Colombian insects eat pink and red Australian *Lantanas*, but not the common pink form from South Africa; they like the red Colombian *Lantana* best. Some insects control the plant by stripping the plant of its leaves and flowers. There is no certain method of eradicating these pesky plants, though the search goes on in Brazil and Central America for an insect that will help rid the world of this manmade scourge.

Crowding in Florida

The state of Florida is one of the major points of entry for plant materials coming into the continental United States. Altogether, roughly 170 species of ferns and higher plants have established themselves in Florida, disturbing the remnants of the natural ecology of southeastern Florida. Some newer additions, such as *Casuarina,* an Australian tree that resembles a pine tree, are so prolific on sandy banks and beaches that they prevent sea turtles and alligators from finding sufficient resting sites.

Some scientists have calculated that about 1,800 exotic plant species have escaped into habitats in North America. Many of the plants that we assume are native plants are not natives at all. In some areas, more than 10 percent of the species are exotics. In Southern Florida 16 percent are escapees from other areas.

One of the first foreign trees to cause trouble in Florida was the Brazilian pepper tree, *Schinus terebinthifolius.* As early as 1898, seedlings were brought into Florida and distributed over the southern portion of the state. By the end of the 1950s, the trees, which produce clusters of attractive red fruits that appeal to mockingbirds and robins alike, appeared in nearly all the different Floridian plant habitats. Wherever it grows, it inhibits the growth of native species.

Perhaps there should be a special hall of infamy for people who have fostered ecological disasters unwittingly or wittingly. One such disaster—at least for Florida—involved Australia's Cajeput tree, *Melaleuca quinquenervia.* The tree's history in Florida began with John C. Gifford, a man of many trades: a forester at the University of Miami, a land developer, a banker, and an entrepreneur. Gifford let loose the Cajeput tree in 1906. Soon the species began to invade the coastal marshes and boggy grasslands of Florida. As late as 1936, Hully Sterling, Gifford's colleague, flew over the everglades and scattered Cajeput seeds. We do not know for sure if Gifford and Sterling connived to reforest the everglades, but we do know that Cajeputs sprouted, reseeded and multiplied. They formed inpenetrable thickets that crowded out most natural plants and were shunned by wild animals. The everglades now contain hundreds and hundreds of acres of Cajeput forests. In these areas, the native species have been trimmed down 20 to 40 percent from their original number.

The Cajeput and pepper trees both cause skin rashes in allergy-prone people. Perhaps these poisons have helped make these trees so successful. Herbivores detect these poisons and probably avoid the trees. So far, no control methods are known.

Creeping Cacti

Prickly pear cacti are American plants that have caused many problems and heartaches around the world. A number of different species exist, but only a few have been successfully controlled. These cacti belong to the genus *Opuntia,* and they come in one of two forms: one type has green flattened, circular pads with tufts of thorns; the other type has cylindrical jointed stems. We will refer to the flattened ones as prickly pears and the cylindrical ones as jointed cactus.

Prickly pears are originally Central American oddities that were brought into cultivation and disbursed around the world more than two centuries ago. One species, *Opuntia ficus-indica,* grew accustomed to the drier regions of South Africa and formed impenetrable thickets there that rose 12 to 15 feet high. The prickly plants grew so thickly that they crowded out the natural vegetation and prevented agricultural activities. Nearly one million hectares have been infected with this species of the prickly pear.

The prickly pear may be annoying now, but it was not always a noxious weed. For the first hundred years it was cultivated as an ornamental or for its tasty fruit. Just before the middle of the nineteenth century, however, the plant began to expand its territory, and by 1891 it had infested more than 300,000 hectares of land. When the South Africans became aware of the weedy growth of the species, they began exploring ways of wiping out the pest. First, they tried to mechanically clear the areas by cutting the plants out of the soil. The broken pads just re-rooted. South Africans found that they could not keep up with the rapid re-infestation rates. The next attempt was to try chemical herbicides based on arsenic. These herbicides provided limited success and were used up until the middle of the present century.

Help eventually came from Australians who were battling two other prickly pear species, *Opuntia inermis* and *Opuntia stricta,* both of which had also become noxious weeds. During the first part of this century, the Australians searched for an insect that could demolish the prickly pear. They first tried a phyticid moth without success. Another moth, *Cactoblastis cactorum,* was able to devour *Opuntias* in America where the cacti naturally occurred, but these moths initially died out in Australia. When the moth finally took hold, within twelve years the winged creatures had cleared 27 million hectares of the two *Opuntia* species.

Upon hearing of the Australians' success, the South Africans sent for the *Cactoblastis* moths, bred them and then released them into the prickly pear-infested countryside. At first it looked as if the moths would work. The cactus was damaged by the moths' attack, but the plants kept responding from the woody base. The South African plants were obviously going to be more difficult than their Australian relatives. Several other insects were imported. These seemed to be effective but other insects successfully preyed upon them. Two wood boring insects also were released; these proved to be only partially useful. Nothing has proven to be a sure thing against the South African *Opuntias,* but the combined efforts of the insects have at least reduced the rampant infestation in the country. About 75 percent of the originally infected area has now been cleared out.

One other *Opuntia* species (*Opuntia aurantiaca*) unfortunately exists in South Africa. The jointed cactus will not yield as easily. The plant's limbs break off at the joints, and roll along the ground to root where they come to rest. The plant is reminiscent of the dreaded Hydra of Greek mythology—the more the plant is smashed, the more plants result. *Cochineal* moths, and all other insects that have been tried, are ineffective at controlling this *Opuntia*. Similar to *Lantana*, people suspect that this *Opuntia* is a hybrid plant on the loose, since the plants are almost sterile and produce very little seed. They propagate almost exclusively by fragmenting. The cactus' large, wicked spines easily catch in the fur or wool of passing animals. This helps spread the plant.

An Ecological Curse

Another ecological curse is nicknamed just that: Curse. *Clidemia hirta* is a dense tropical shrub, nearly twelve feet tall, with hairy stems that bear prominently veined four-inch leaves. The pinkish flowers are followed by sweet, dark-blue berries. This plant sounds attractive, but it is another aggressive, invasive plant that swallows space at the expense of natural vegetation. The plant, called Koster's Curse or simply Curse, is most prominent on islands of the IndoPacific region, from Madagascar and Sri Lanka through Indonesia to Fiji Samoa and up to Hawaii. The Curse invades both sunny fields and the shady floors of rubber and other tree plantations.

When found in its home territory—South and Central America and the Caribbean Islands—Curse is hardly a curse at all. It is found at the edges of clearings and in moist surroundings; it also is found in cocoa plantations in the Americas. In these native habitats, the plant is not considered a serious weed or pest. Why is Curse so docile in its home range but so aggressively invasive elsewhere?

Curse was brought to Fiji before 1890. By 1920, the species had become widespread and already was impeding normal agriculture. Curse bushes rapidly filled valleys and agricultural districts. The alarmed islanders looked for reasons why the plant was restrained in its natural habitat in the Americas. They discovered a tiny, sucking insect—a thrip—that fed exclusively on *C. hirta.* American thrips were imported and released in Fiji where the insects flourished and started to demolish the curse. Unfortunately the thrips only fed on the plants in sunny clearings, not the Curse plants inside the Fijian forests. The thrips apparently avoid shady places so the problem is only partially solved. Scientists continue to search for natural weapons—insect pests—to battle against Curse. So far no other satisfactory species have been found.

Meanwhile, the Curse has spread to even more islands. *C. hirta* was first reported on the island of Oahu in Hawaii in 1941. Eleven years later a dense stand of *C. hirta* shrubs covered 240 acres. In 1954, authorities in Hawaii released the thrips onto the island. Sure enough, the insects cleared up the Curse plants growing in the sunlight. The plants growing in the forest were again unaffected. Curse currently grows on about 77,500 acres of Oahu forests and is so dense that it accounts for half of the vegetation on 20,000 of those acres.

In 1972, the plant had spread to the Big Island, Hawaii, and natives noticed it on the island of Molokai the following year. Despite efforts to curb the growth of these plants, Curse has maintained a toehold on these islands.

On the Hawaiian islands, and other islands too, the first appearance of Curse triggered concern for farm lands. When thrips controlled Curse plants in these open areas, little attention was given to Curse in the forests. The short-sightedness of this approach is apparent now as we watch native species disappear.

Kudzu

Just as the tropical islands have their Curse, the southeastern states of the United States have Kudzu, a spectacularly rampant climbing plant. Kudzu, actually *Pueraria lobata,* a member of the bean family, comes from Japan and was originally brought into the United States as an ornamental and a possible source of cattle fodder. It has quite large leaves, with two or three lobes. Since it is a legume, you would expect it to be quite nutritious. However Kudzu has never really caught on. In what turned out to be a very poor idea, some people in the United States decided to use Kudzu to control soil erosion. In the 1930s, the Civilian Conservation Corps (CCC) planted thousands of the plants across the South, and so the nightmare began. The plants enjoy the South's long growing season, the rainfall, and the high summer humidity. The vines can grow more than 75 feet in a single season and, depending on conditions, will reproduce by seed or vegetatively. When a creeper stem touches the ground, it can root at that spot and proceed to make new plants and new climbing stems. Kudzu plants are rampant climbers, spreading over whatever provides support, whether it is a tree or a house. Kudzu is so dense that the plants keep light from reaching the trees beneath. Without sufficient sunlight the trees die. Chemical herbicides are used to combat this tenacious weed, but no effective controls have been perfected.

Green Cancer

Nowhere are weeds more devastating than South Africa. The South African Cape vegetation has been reduced to less than 36 percent of its original area. Those plants not threatened by urban expansion or farmland conversions are often attacked by alien weedy trees and shrubs. These aliens will probably crowd out the remnants of the natural ecosystem. Among the plants posing a specific threat are the various species of *Acacia,* many of which enjoy sandy flatlands, and *Hakea,* which grow so rampantly on mountainsides that they have been called Green Cancer. Both are originally from Australia. In all, 28 different species of weed are causing trouble at the Cape.

Eight species of *Acacia* were introduced to South Africa—some to stabilize sand dunes and others for timber and tanning bark. *Acacias* also occur naturally in Africa, but insects eat more than 90 percent of the seed from the African *Acacia* species. None of the African insects will eat the seed of the Australian aliens; so the foreign plants have spread aggressively, overshadowing native plants. Insects imported from Australia eat only 15 to 25 percent of the seed production. The control problem is complicated by the fact that *Acacia* seeds can stay dormant in the soil for about forty years before sprouting.

Hakea is one of the nastiest weeds at the Cape. Not only does it aggressively crowd out other plants, but each of its cylindrical leaves is tipped with a sharp spine. It is difficult to handle or cut the shrub's tough stems without being punctured by these needles. Mechanically cutting the brush on steep mountain slopes is like slashing at a thousand steel needles with a machete: The machete will win but the needles draw blood.

Besides the weedy *Acacias* and *Hakeas,* pine trees have also become invasive weedy trees at the Cape. Pines set quick growing seedlings—shading the native plants and smothering the ground with a blanket of dead pine needles. Both seed germination and the growth of small plants are prevented.

The Perfect Weed

A grass from the Pampas of Argentina and Uruguay, first called *Nassella trichotoma* and later *Stipa trichotoma,* is bad news. *Nasella* has been described as the perfect weed, being extremely efficient and successful. A single plant will produce 100,000 seeds in one season. Most of the seeds live for just five to seven years, but about 5 percent will live at least a couple of decades. Once *Nassella* invades an area, a twenty-year program usually is needed to rid the area of the pest. Mature plants can be killed but seedlings continue to germinate each year. The process appears to be never-ending. New Zealanders spend millions of dollars each year on this cycle of *Nasella* control. The grass is not spectacular in appearance and does not attract attention. Plants creep into areas without tipping off their arrival until it is too late. Almost without warning, the land is covered with useless, non-nutritious grass. *Nassella* tolerates a wide variety of conditions—acid to alkali soils, moist to dry climates, and areas with either summer or winter rainfall. New Zealanders have found that one way to deal with the grass is to smother it in pine forests since it does not like heavy shading. A pine plantation takes ten years to completely smother *Nassella,* and this method is useful only if you want pine plantations. When *Nassella* wipes out natural vegetation, replacing the grass with pine trees will not bring back the natural plant species.

Gladiolus aureus

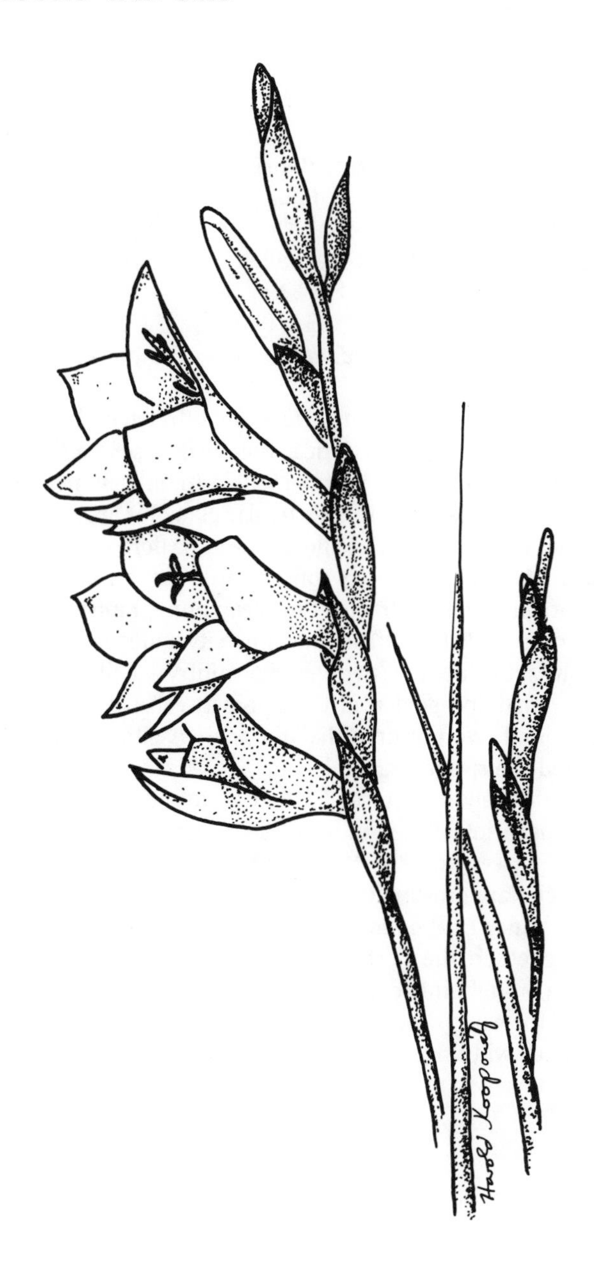

Case History: The Golden Gladiolus

Years ago, before the extinction crisis penetrated so deeply into the flora of South Africa, meadows of short plants flourished along the flat places at the base of hills. In winter, when it drizzled many days in a row, bulbs and corms tucked beneath the soil began to sprout. One gladiolus plant, a slender wand with four or five drops of molten gold, wafted in the breeze, mimicking the African sun above.

Many of the numerous gladiolus species found on the old continent are on the endangered lists. Of these, *Gladiolus aureus,* the golden gladiolus, has received the most attention.

This rare species occurred only along a four-mile stretch of the western coast of the Cape Peninsula in southern Africa. Botanists observed the number of individuals plummet from seventy plants in 1975 to only eighteen plants two years later. The problems faced by this particular gladiolus are shared by nearly 1,500 other species of threatened Cape flora. The Cape flora is one of the richest in the world with about 6,000 species concentrated at the continent's tip. The even and pleasant climate has attracted European immigrants since the founding of Cape Town in 1652. Today, the sprawling city and its suburbs usurp the land. Smaller towns, too, dot the surrounding countryside like moths attracted to a flame. The remaining golden gladioli lie between picnic grounds and a children's playground. We presume that the many visitors who have picked the blooms caused the species' rapid decline.

People are not the only reason this flower has not flourished. Foreign trees and shrubs have been introduced into this same area and are now spreading out of control. Pine trees from the Mediterranean, acacias from Australia, and other foreign species are growing up in thickets and choking the slender flora of the Cape. The spiney *Hakea,* considered a green cancer, is worst of all.

Nearby at Kirstenbosch, the most famous botanic garden in Africa, the golden gladiolus has been brought into cultivation and seed has been sent to interested gardeners around the world. While its prognosis in the wild is poor, we suspect the golden gladiolus will linger for a while in cultivation.

III.

A TECHNOLOGICAL ARK

CHAPTER 12

How to Make Your Own Gene Bank

One practical and effective method to combat plant extinctions is the gene bank, a freezer of sorts that preserves seed and pollen. Creating a "technological ark" does not require extensive knowledge or specialized training, nor does it require expensive equipment. The simplicity of a gene bank makes it a viable project for a native plant society, a plant hobby group, or even individuals. Anyone who wants to make a personal contribution to the plant conservation effort can do so in their own garage or backyard. In this chapter we will explain the basic theory and hands-on mechanics of gene banks. Gene banks generally are easy to construct and maintain, although a few problems may arise during or after construction of the bank. The following pages offer guidance and warnings to help new gene bankers sidestep such mishaps.

Many people mistakenly assume that governmental agencies or educational institutions are the most appropriate groups to operate gene banks. Actually,these groups are unreliable over the long haul because their funding priorities are not permanent. Government funding is regulated by politicians who may or may not be concerned with saving plant species.

Reliable gene bank operators often are plant hobbyists, who not only are devoted to furthering their preferred group of plants, but are also interested in conservation. Species collected eagerly by plant hobbyists may receive minimal attention at arboreta and botanic gardens. Furthermore, changes of arboretum directors may signal a change in emphasis on which plants are to be saved. Hobby groups, therefore, can have a powerful effect on the effort to preserve plants. Some groups, such as those that breed and cultivate ferns, cacti, and alpine plants, already operate some form of a seed or spore bank. For such groups, creation of an actual cryogenic gene bank would simply extend their preservation techniques.

The ABC's of Gene Banking

The theory behind gene banks really is quite simple. Cryogenic preservation is no more than suspended animation at subfreezing temperatures. While this may conjure up visions of sophisticated high technology, the process requires little more than a household chest freezer; storage containers made of glass, metal, or plastic; a simple desiccant or drying agent to dehydrate the seed; and the seeds themselves.

Seeds are tiny organisms that stay alive so long as there is a food reserve inside the seed to fuel vital chemical reactions that keep the seed alive. These chemical reaction rates are dependent on the surrounding temperature—the hotter the temperature, the faster the reactions will take place and the less time the food reserves will last. The converse also is true. For every 5°C drop in temperature, the life of the seed will double. For example, onion seeds with 10 percent moisture are good for 16 weeks at 35°C but will live for 78 years at 0°C. Dropping the temperature down to -15°C would increase longevity to 624 years, provided the ice crystals caused by the 10 percent moisture do not harm the seed when frozen. Clearly, if seeds could be stored at subfreezing temperatures, they might be good for centuries.

Given this information, how cold is cold enough? Two factors dictate the storage temperature: the desired cost of the freezer and the desired longevity of the seed. The freezer cost increases as the temperature inside the apparatus decreases. Figuring desired longevity is not so simple. Cryogenics theory suggests that extremely low temperatures allow seeds to last for many centuries, but such tremendous lifespans may not be necessary. If human beings somehow manage to persist through the next century or two, we suspect they will have come to terms with their planet's ecological problems. One or two hundred years of storage, therefore, probably is sufficient.

Besides the storage temperature, the other major concern is water. Too much water in a seed will form ice crystals as the temperature dips below the freezing level. Seeds with a high water content are plagued by large ice crystals that rupture individual cell membranes and destroy the seed. Many seeds have a naturally reduced water content and can be frozen soon after they have ripened without danger of rupture. Some seeds, however, require extra care to reduce their water content. While this usually is a fairly easy task, species with extremely fleshy or oily seeds simply cannot be frozen and must be preserved by other methods.

Just how much water should be reduced? The life of a seed will double with each 1 percent decrease in water content. A minimum of 4 percent water content appears to be required by seeds to stay alive. International standards for longterm seed storage suggest that an average seed should contain 5 percent water and be stored at −18°C. When both temperature and water content are reduced, the two factors multiply—when temperature drops 5°C and water decreases by 1 percent, the seed lives four times as long.

Collecting, Cleaning and Drying

The first step in collecting for the gene bank is to make certain that the seeds are alive. A variety of biochemical tests and stains can be used, but the most dependable method is also the easiest. Simply plant a known number of seed and see how many germinate. Only on rare occasions will all of the seeds begin to grow. Plant scientists usually strive for about 95 percent germination. Some plants have naturally low germination rates. For example, we have found that *Aloe albida* germinates at a rate between 25 and 30 percent. This species is endangered, perhaps because of this highly inefficient rate of germination.

Because space in a cryogenic gene bank usually is at a premium, don't waste space by storing chaff, pieces of seed capsules and stalks, or dead seed. Cleaned seed takes up less space in the bank and is easier to inspect and observe. At our gene bank in Irvine, we usually harvest pods as soon as they start to ripen or split, put the seeds in a container, and expose them to air for a few weeks. When first extracted from pods, the water content in seeds is quite high. However, Southern California's low humidity allows the seeds to mature and dry in the air. This occurs because seeds match their moisture content with the water vapor in the air. They either lose or absorb water, depending on whether the water vapor in the air is higher or lower than the water in the seed. Dry air results in dry seed. Air-dried seeds harden and can be handled safely. The National Seed Storage Laboratory in Colorado also benefits from low humidity and contains mostly air-dried seeds.

Seeds can be dried simply in humid areas by storing them for several days in a closed container with a desiccant or drying agent. The most familiar desiccant is *Silica Gel,* a product widely available in hardware, hobby, and drug stores. Another product which readily absorbs moisture from the air is *Drierite®* a calcium salt. *Drierite®*, comes with a color indicator: when it turns blue, it can still absorb moisture; when it turns pink, it is hydrated and can absorb no more. One word of caution—only a fresh desiccant can absorb moisture.

It is important to test the desiccant before using it to dry seed. A handy test for freshness is to dip a strip of filter or blotting paper in a saturated solution of cobalt chloride. When the paper dries, it will turn blue if the air is dry, and pink if the air is moist. When you discover that the desiccant is stale, rejuvenate it by baking it on a teflon-coated cookie sheet for two or three hours at 60°C or higher. This will dry it back to its original state.

Now that the desiccant is ready to use, find any container that can be sealed. An inch of desiccant should be placed on the bottom of the container and covered with either cheesecloth or wire gauze, permitting air to reach the drying agent. Open containers holding the seeds should be put into the desiccator and the desiccator sealed. Open the desiccator at least once a day to stir the seed. During the first two days the seeds will lose most of the water content. A little more is lost during the next five or six days. Within a week, most seeds can be safely frozen. This is the most practical method of drying seed for a home gene bank.

How Dry is Too Dry?

There can be too much of a good thing. Overdried seed can result if proper care is not taken. Some botanists recommend air-drying at 35°C for several days. This process has been used routinely to dry many different kinds of seed, but we always have been rather wary of unknown longterm effects. During the drying period the seeds are subjected to considerable temperature stress, and an acceleration of the metabolic processes. In our opinion, drying seed at the gentler room temperature is desirable.

People frequently ask about freeze-drying, a process that rapidly freezes tissue and evaporates most, if not all, of the water. Seeds subjected to freeze-drying probably would die due to considerable ice crystal damage and the strong vacuum used to dry the material. There are reports, however, that pollen has been stored successfully following freeze-drying treatment. This is a prime area for further research. For seed, we advocate the safe two-step process of first drying and then freezing.

Storage Containers

There is just one cardinal rule about storage containers: they must be absolutely airtight so that already dried seeds will not take up water from the air. The air within a freezer contains water molecules which attach to dried seed if the container is not airtight. Doubters can place a piece of cobalt indicator paper into a freezer and see the results of the moisture. The ice buildup in a freezer is further proof.

Many kinds of containers can be used to store seed, but the best are those with seals caused by the fusion of the material in the container. Three varieties—metal, glass, and plastic—appear to be almost foolproof. Metal cans probably are the sturdiest containers, but without a machine to weld the lids into place, these are both troublesome and expensive. Large samples of big seeds (such as beans and peas) store very well in metal. Glass containers, the second option, are more fragile but this method is relatively cheap. This is what we use in Irvine. Inexpensive disposable glass test tubes are easy to get from scientific or medical supply houses. We seal the end of the tube with an inexpensive propane torch, found at most hardware stores. With a little practice the tube can be sealed within a few seconds without significantly heating up the seed sample. A handy feature with glass is that it is easy to view the frozen sample. The third type of container is an envelope made of foil-laminated plastic that can be heat-sealed. These envelopes have an inner lining of polyethylene and an outer lining of aluminium or some other metal. While the polyethylene is slightly permeable to gasses and could allow some water molecules to cross into the envelope, the metal lining is impermeable and insures that the container will be airtight. The plastic will melt and seal when heat is applied to the open end. You can add a vacuum device to suck most of the air out of the package. These envelopes take up less space and are easiest to store, but they are not easy to find.

Labeling and Access

Imagine that you have a freezer full of packages, tubes or cans of seed and that you must retrieve a specific packet. This is one of the most difficult problems associated with small gene banks. Bankers need to figure out the most efficient way of storing large quantities of small packets so that one can be retrieved without defrosting the entire collection. A system of numbered shelves within the freezer seems to be the most practical way around the problem. Drawers are impractical and should be avoided because ice buildup could freeze the drawers shut. It is important to label all samples clearly because the label must resist subfreezing temperatures for long periods of time. You must use ink that will not fade or become brittle, and the paper must not self-destruct. We believe that a dual system of labels probably is the best bet. One label should be placed on the outside of the sample and another should be placed inside. As an extra precaution, a record of the sample's shelf position should be recorded in a book or mini-computer. The name of the sample should be imprinted onto a metal foil label. This is one of the best possible longterm labels. The internal label is the most important one as it will always accompany the seeds. It also is a good idea to number the samples. This accession number could be correlated with the year of the sample and stored with the permanent record.

Record-keeping obviously is a vital part of operating a gene bank. Records should be precise and document the quality and viability of the seed. The original source of the seed and the date of processing are further useful bits of information to retain. The more complete the record, the more valuable the sample. Notes on other parameters would be useful, but don't get overwhelmed by records. Seeds should be the prime focus of the bank, not paperwork. In the final analysis, the last seed of a species is more important, even without records, than a very fine file about an extinct species.

How Large a Seed Sample?

There is no simple formula to determine how large a seed sample should be. The size should depend on the ultimate use of the sample. If the sample is needed to conserve the entire gene pool, that is, the full range of variation within the population, then the sample should contain nearly 10,000 individuals or seeds. If only a representative sample is needed to retrieve a few plants for illustrative specimens, breeding experiments, or even genetic examination, small samples of twenty to one hundred seeds might be sufficient. Because the seeds are stored in sealed containers, samples should be easy to extract. Removing a small sample directly from the freezer is better than taking out a large sample, defrosting it to remove the needed amount, and resealing and refreezing the remainder. Seeds can handle some defrostings and refreezings, but we don't know how many times they can suffer this treatment. It is more prudent to save defrostings for brownouts and other occasional power failures.

Ideally, the sample of a particular species should have several components. First, it should contain a "basic" foundation sample that should not be touched except in dire circumstances. Second, it should hold another large sample that can be used to search for a particular genetic variant. Finally, smaller samples should be reserved to fulfill requests for seeds from scientists, institutions, or their individuals who might wish to examine representative plants or products.

Because gene banks are susceptible to the variety of accidents and catastrophes that can affect any institution or building, gene bankers should distribute duplicate samples to other gene banks. Fire, earthquakes, tornadoes or floods could devastate a collection of hundreds of species. Having backup copies in other collections, and allocating a portion of your own freezer for the maintenance of others' duplicates, is a sensible approach.

The Freezer

A few points should be made regarding the type of freezer to use. Upright freezers are a poor idea because cold air rushes out when the door is opened. Chest freezers are the best choice for small gene banks. These range in price depending on the desired temperature, the highest-priced freezers having the lowest temperatures. Freezers with temperatures below −18°C are unnecessary since the longevity of the seed already is longer than the working life of the freezer. We have found that a freezer with a temperature range of −35°C can be purchased for several thousand dollars. Actually, even a regular household chest freezer with a temperature of −15°C is adequate for a gene bank. Technology and science are still unable to prevent ice buildup; the best way we know to handle this problem is to chip away the ice at regularly scheduled intervals.

An important addition to the freezer is a battery-powered alarm connected to the freezer's thermostat that indicates when temperature rises above acceptable levels. Such an alarm is invaluable in signaling brownouts or other technical problems. Another good precaution is a small, portable, gas-driven generator—kept in good working order—that could be used during longterm brownouts.

Pollen and Spore Banks

Seeds are the most likely candidates for gene banks, but this cryogenic technique may be useful in preserving pollen from flowering or conebearing plants, and spores from non-flowering plants such as ferns and mosses as well. People sometimes ask why we bother with pollen, now that the cryogenic technology has been worked out for seeds. One reason is that pollen grains are so tiny that thousands and even millions of grains can be stored in a very small vial. Obtaining great diversity is quite simple. Botanists are able to use pollen grains from certain species to grow whole plants. This still is a difficult, tedious procedure with most species, but we anticipate that future plant scientists will routinely perform these transformations. Pollen grains can also be used when needed to help create hybrids. When seed is used to develop hybrids, the seeds must be retrieved from the bank and then planted. It takes several years of maturation for the plant to be useful in hybridization. Because pollen can be used "as is," it is much more economical and efficient. A pollen bank can be an extremely powerful tool in plant breeding since it frees breeders from the tyranny of time. The process can be streamlined and quickened by crossing pollen from plants that normally flower at the end of a season onto flowers that appear at the beginning of the season.

In Irvine we have found another use for pollen banks. Several specimens we grow at the arboretum are self-sterile—they set seed only if pollinated with pollen from another individual. We had managed to raise *Cyrtanthus obliquus,* a relatively rare African amaryllid, to maturity. This process took ten years and produced five plants, all of which became virus-infected. We decided to replace the diseased plants and also increase our stock from seed. Seedlings are usually virus-free. Here we ran into trouble. Each of the five plants flowered but never two at the same time. Even though each plant produced abundant pollen, none of the plants would set seed from its own pollen. We finally hit upon the perfect solution: storing pollen from this species in the bank. We now have no trouble pollinating the plants, and we can generate as many seedlings as we require. The original five plants were replaced by several hundred offspring.

Fern spores, which are very similar in size to pollen, can be treated just like pollen. Even the spores of tropical ferns like Staghorns—*Platycerium species*—can be germinated after drying and freezing.

The biggest problem with pollen is its susceptibility to water damage. Rain, heavy dew, or water from any other source that comes into contact with pollen is liable to kill it. Pollen should be collected only from fresh flowers that have not been overly exposed to the elements.

We store pollen in gelatin capsules, the type used to hold medicine. These capsules can be purchased from the local pharmacist. The stamen, which contains pollen, is plucked off the flower, inserted into the capsule, and shaken several times to deposit the dust-like pollen on the capsule wall. The stamen then is withdrawn. The species name and date may be written directly on the capsule with a waterproof laundry marker. We dry the pollen by exposing the capsules to the air inside a frost-free refrigerator for 24 hours. The moderately dry air circulating in these refrigerators will cause sufficient water loss to permit the capsules to be frozen safely. Water is able to move out through the walls of a gelatin capsule. We store the capsule inside individual plastic containers that have a little *Drierite*® in them to prevent moisture from moving through the gelatin, and then place these containers in the bank. Relevant data are written on the wall of the plastic container. Pollen samples withdrawn from the freezer have limited viability and should be used within two days of withdrawal. Little is presently known about the longevity of pollen in cryogenic storage. We have tested pollen that had been frozen for three years and found it to be viable, but no one knows yet whether pollen will be as hardy as seed seems to be.

The data presented here and elsewhere on the longevity of seeds are extrapolation figures obtained by keeping large samples at different temperatures, withdrawing seeds and germinating them in successive years, and noting the decline in viability with each test. After a few years the rate of the decline and a forecast of the percent of the seeds that will stay alive at any particular time can be determined. These studies have brought forth a few adverse factors. A main problem is that with time, mutations and chromosomal abberations accumulate. These probably are caused by interactions between the background cosmic radiation, which is everywhere, and the genetic material in the cells of the seed. Since the background radiation appears to be more or less constant, the probability of a mutation occurring depends on the amount of time the seed is exposed to the radiation. As the seed's longevity is increased, so is the chance that a mutation will appear.

Other longterm effects can also occur. The proteins in the seed may degenerate. Also, despite the low temperatures, some chemical reactions still occur. Finally, molecular events that we know nothing about may take place. The effects of these changes do not appear on our extrapolation curves. Nevertheless, we are totally convinced that cryogenic banking is presently the most effective and efficient way to preserve species.

As we mentioned earlier, not all seed can be stored cryogenically. Some plants have fleshy seeds with high water content. The seeds of some Crinums, for example, resemble small potatoes and simply cannot be dried. Nerine, another related plant, has seeds that begin to germinate before they are released from the mother plant. Dehydration would kill the seed. Seeds with very hard seedcoats may not be dryable either, and seeds with very high oil content can stubbornly resist drying. These are isolated problems; most species can be processed for cryogenic storage. Further experimentation should show us how to deal with some of the fleshy and oil seed. Viability of seed in storage may also be affected by the amount of oxygen or other gasses available, but these effects and other relatively negligible factors shouldn't concern the beginning gene banker.

Despite what seems to be a complicated series of steps, anyone with determination and a little effort can create a gene bank. Extensive know-how or technology culled from futuristic science fiction is not necessary. Gene banking is an elegantly simple solution to a vexing and important problem.

Moraea neopavonia

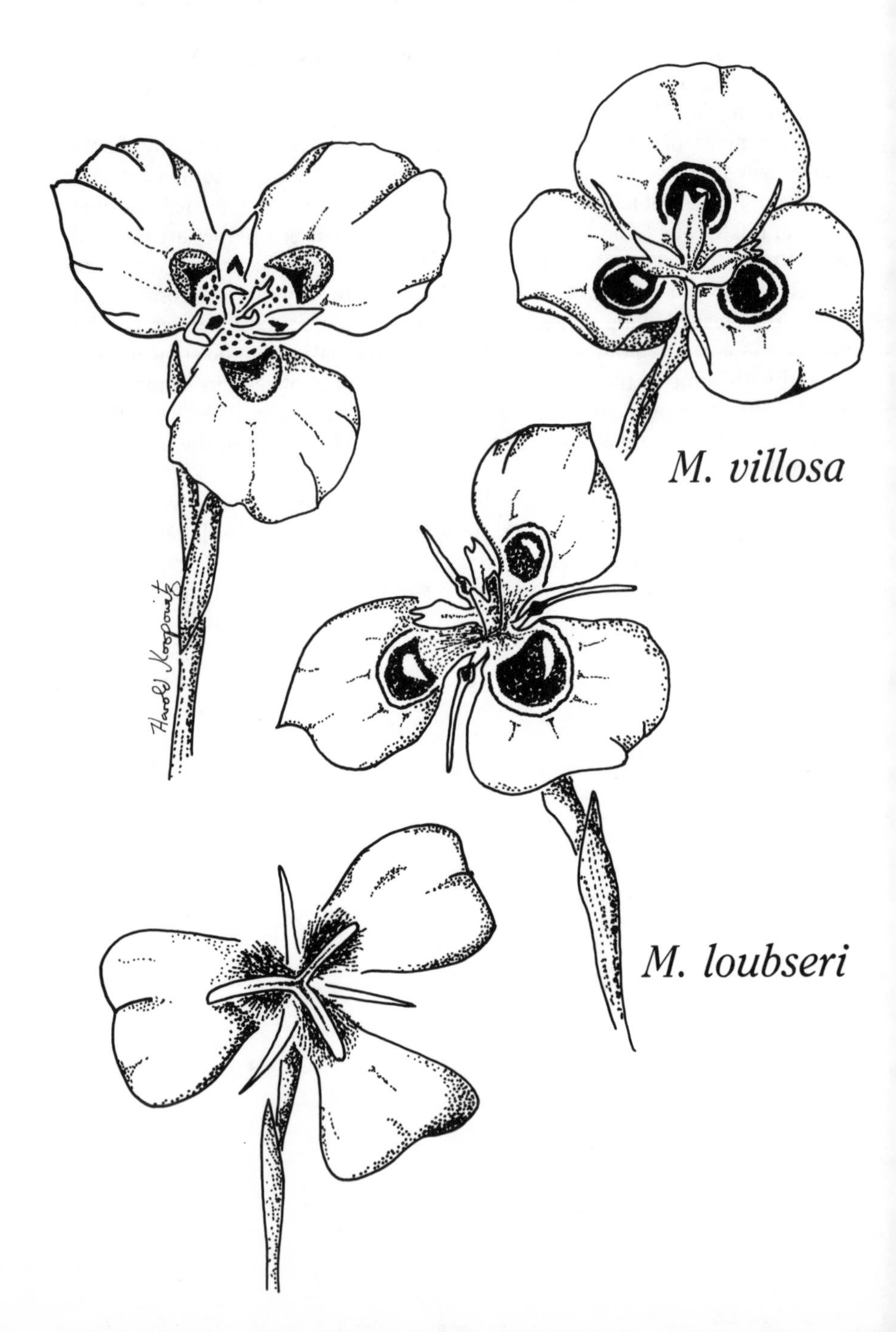

Case History: The Peacock Moraeas

Moraeas are the Southern Hemisphere's answer to the Iris. Some moraea species closely resemble garden-variety irises found throughout the United States. A few botanical traits differentiate the two but, overall, these are relatively minor discrepancies. Many moraeas have bright, colorful flowers that dance above grassy foliage like gently hovering butterflies. A few species are as ephemeral as the wind—their pale, dusky flowers blooming in the late afternoon sun for just one hour before gently folding up at nightfall. Other robust species produce clouds of vivid blossoms that live up to three days. When these flowers fade, others quickly take their place, creating the effect of almost perpetual bloom. One species, *Moraea polystachia,* has such dense blooms that from a distance the blue flowers glisten like pools of water. This species is not yet endangered, but other moraeas are not as lucky.

This case history concerns a group of moraea species commonly referred to as the peacock moraeas—a spectacular group having three large, rounded petals, and three much smaller, narrow petals on each flower. An iridescent blue or green spot marks the base of each large petal and is surrounded by contrasting hues, resembling the shimmery markings on a peacock's feathers. A few peacock moraeas are still plentiful but at least four species appear to be headed toward total extinction. One of the four, *Moraea amissa,* has not been seen for many years and some people fear that this species already is extinct.

Another species, *Moraea aristata,* a white-petaled, blue-eyed member of the group, was widely cultivated in Europe under the name *Moraea glaucopsis* and occasionally still is seen under that name. This species was widely grown in Europe during the mid-eighteenth century—its elegant flowers attracting plant collectors who visited its home territory in Africa. Even in the United States the species became popular. Gardening books written during the first part of this century indicate the species was readily available. As this century moves toward its close, however, *Moraea aristata* is becoming increasingly rare. This plant commands exorbitant prices when available. Some scientists believe there are more dried and pressed specimens of *Moraea aristata* in museums and herbaria than living in the wild. What happened to this once common species?

Because we still do not know the original sources of cultivated *Moraea aristata,* it is difficult to fully understand the problem. Perhaps the main source was wild-collected bulbs which were finally depleted. Or maybe the Dutch nurseries that used to grow these plants found it was no longer profitable to continue the cultivation. Another possibility is that the cultivated stocks were drastically reduced during World War II. Whatever the reason, this example clearly shows that despite abundant numbers in cultivation, the species has not been saved forever. *Moraea aristata* now grows wild in just one small area. Fortunately this patch of rare flowers grows in an undetected cite near Cape Town where it is protected against both poaching and encroachment by developers. Even this protection does not insure safety since the few plants that remain in the wild population produce only a few flowers. The chances are remote that two flowers will appear at the same time so that they can be cross-pollinated and set seed. The amount of seed produced by these frail plants is substantially less than what is needed to balance the normal deaths of mature plants. You do not have to be an expert to see that this species is on the way out unless herculean efforts are made.

Even more beautiful is *Moraea neopavonia,* a close relative of *Moraea aristata,* with bright orange flowers that contain glistening peacock dots. This species apparently never was very common. Because it grows in prime farming country, it has been reduced to a few localities. No one knows if *M. neopavonia* can survive the ever-increasing need for agricultural land.

A fourth peacock, *Moraea loubseri,* was discovered in 1973 on the summit of a small hill in the southwestern Cape of South Africa. It has prominent blue or violet flowers but, instead of having shiny spots in the center, each petal has a dense tuft of black hair. The flower seems to have furry bumble bees sitting on its petals. *Moraea loubseri's* habitat no longer exists. The hill became a quarry, used to supply crushed rock to build man's ever-burgeoning artifacts. Even though the habitat was totally destroyed, the species was not lost. A few specimens of *Moraea loubseri* were given to the man for whom the species is named, and because seed from a few plants was widely distributed, the species managed to survive and now lives in captivity in a number of locales, including the arboretum at Irvine. Even so, we shouldn't forget that captive populations are never totally safe.

CHAPTER 13

Other Banks

Since cyrogenic gene banks are relatively easy and inexpensive to operate, why are there so few? Many people do not realize that cryogenic banks involve such simple techniques. For them, cryogenics conjures up images of technicians in white coats wheeling gleaming vats of liquid nitrogen. As the simplicity of these banks becomes widely known, more and more facilities will be developed. Already, some of the institutions and groups that currently use gene banks without subfreezing temperatures have been persuaded to adopt the more efficient, cryogenic techniques.

The value of gene banking was first discovered in the late 1960s and early 1970s, when plant scientists began looking into the possibility of saving the genetic variability in agricultural crops. The concept was recently expanded to employ subfreezing temperatures and include non-agricultural plants. In this chapter we'll take a look at the history of gene banking and at some of the different gene banks around the world.

Natural Gene Banks

Gene banking may be a relatively new concept for mankind, but natural gene banks are a regular part of nature. These natural ones lend support to the theory that seeds preserved today in manmade banks will be viable hundreds, if not thousands of years into the future. Soil is a kind of gene bank to which seeds are added year after year. When wild plants produce seed each season, not all of the seeds will grow the following year. Much of the seed drops to the ground and then waits for appropriate conditions to permit germination. Most seed gets covered by either dirt or dust and eventually is buried. The number of seeds that accumulate over the years can be prodigious, as seen in a study made in Denmark. One block of topsoil—one meter square and twenty centimeters deep—yielded roughly 135,000 seeds. Of these, about 50,000 were living and could germinate under the correct conditions. Compared to other studies, this is an extreme example, but it is not unusual to recover thousands of viable seeds from a square meter of land. If the plants in an area change, then the species in the natural bank also will change. Deep in the soil there may be species no longer common. At the University of California, Irvine Arboretum, we once decided to add a few small terraces and so removed soil from one area to a depth of two feet. Much to our surprise, a field of stinging nettles sprang up after the next rains. Until this time, no nettles had been found on the grounds nor in the surrounding area. The seed must have been buried decades before and was waiting patiently for the right conditions. These natural gene banks often are seen when a forested area is cleared. All sorts of plants suddenly germinate from seed that has been long buried. The plants may be totally unrelated to the normal flora of the forest.

The length of time that seeds remain viable in the natural seed bank depends on a variety of conditions. To begin with, the colder the climate, the better the chances of survival. An interesting study involved seed dug up from beneath Danish churches built centuries ago. The soil beneath one church (1,700 years old) yielded viable seeds from two weeds, *Spergula,* commonly called corn spurry, and *Chenopodium,* or lamb's quarters. Another 600-year-old church had soil laced with thirteen different viable species. Lotus seeds dug up from a peat bog and dated 1,040 years-old have grown, as have the lotus seeds rescued from a 237-year-old herbarium specimen. The record is held by lupine seeds dug up from frozen soil in the Arctic. The seeds were radiocarbon-tested to be about 10,000 years old and germinated when placed under the correct conditions. While some seeds seem to be able to last almost indefinitely, we should remember that there are many species whose seed does not stay viable for more than a few months or years. Nevertheless, the ability of many species to remain viable for centuries validates the theory of artificial gene banking.

Besides the need for cold temperatures, seeds need to be relatively dry to survive in a gene bank—whether it is natural or artificial. Many ripe seeds tend to have seed coats impermeable to water. Before the germination process can begin, the seed coat must break down so that the seed can absorb moisture. Dry seeds are safe from germination and may remain in this state of suspended dormancy for many years.

Geographic Centers of Genetic Diversity

Looking back into history, we see that domestication of crop plants was a tedious process that stretched over centuries. Some crops go back 10,000 years. Vavilov, a Russian agriculturalist, appears to have been the first to discover that some geographic areas were naturally rich in primitive varieties of food plants and their wild relatives. Some of these natural centers were rich in one particular crop and others were abundantly endowed with many primitive crop plants. Turkey and the surrounding lands of the eastern Mediterranean and Near East form one such area.

Over the millennia, groups of plants in these areas had evolved a rich diversity of genes within nature's balance. These plants resisted many diseases and pests, and could survive drought or extreme temperatures. Vavilov pointed out that these centers of genetic diversity were ideal places to hunt for genes needed to enrich agricultural crops. Scientists in the mid-1930s first warned that these genetic centers were disappearing, but the alarm fell on deaf ears. In the early 1960s, the Food and Agricultural Organization (FAO) of the United Nations, and the International Biological Program (IBP) investigated the genetic potential of crop plants, but neither group was concerned with saving non-agricultural genetic materials. However, a growing number of scientists, alarmed by the degradation of these centers of diversity, began to stress the vulnerability and importance of the genetic materials within these regions.

By the early 1970s, the FAO was being urged to organize an international effort aimed at conserving genetic resources. Many individual nations took the first steps. Some countries, including the United States, had already begun to collect and maintain seeds of agriculturally important crops. The United States' National Seed Storage Laboratory (NSSL) at Fort Collins, Colorado was in operation at that time. The Vavilov Institute in Leningrad had enlarged its activities and collections; the Izmir Center was established in Turkey, one of the most important centers of diversity. Other collections in Mexico and the Philippines were created. In 1974, the International Board for Plant Genetic Resources was set up in Rome, under the auspices of the FAO, to oversee conservation of crop materials. A number of scientists kept their voices raised and today the importance of preserving ancient materials is recognized as an important aspect of modern plant breeding, although not universally practiced.

The numbers in some gene bank collections are staggering. The International Rice Research Institute (IRRI) at Los Banos in the Philippines has about 60,000 rice accessions. Some of the rice species have been developed into the high-yielding dwarf rice varieties that now feed much of the world. An important problem for banks this large is the regeneration of seeds from time to time, particularly if they are not being stored at subfreezing temperatures. Regeneration should occur at about 25 to 30-year intervals to ensure viability.

In recent years these various institutes have been switching to cryogenic seed banks, using subfreezing temperatures and partially dried seed. Now, samples can be viable almost indefinitely without needing to be rejuvenated. Gene banks generally use two storage standards: the "preferred standard" storage at −18°C or below and moisture content between 4 and 6 percent, and the more common "acceptable standard," at 5°C and moisture content between 5 and 7 percent. These numbers may sound rather similar, but in terms of longevity they are quite different. Wheat stored at the acceptable standard must be regenerated every fourteen years to remain viable. The preferred standard requires regeneration only every 390 years.

The National Seed Storage Laboratory

Germ plasm or gene bank collections in the United States started when the Department of Agriculture established four Regional Plant Introduction Stations to assess new crops before introducing them to the farming community. Later, an Inter-regional Potato Introduction Station was established. The efforts culminated in the National Seed Storage Laboratory (NSSL) at the University of Colorado in Fort Collins. The NSSL does have some areas for subfreezing storage, but most of the accessions are stored at 5°C. In relation to the significance of agriculture in the American economy, the federal support given to NSSL has been miniscule. On a worldwide basis, only 3 percent of research money goes to agriculture, and the tiniest fraction of that goes to gene banks. In recent years collecting and maintaining diversity within gene bank centers have received more emphasis.

The Vavilov Institute

The primary Russian gene bank is the N.I. Vavilov Institute in Leningrad, with twenty-five experimental stations around the country. The facility houses an enormous collection of seeds and living plants ranging from wild species and primitive crops to breeding lines of modern crops, though most of its work is in plant introductions. In 1975, the institute's collection had more than 200,000 accessions due to its aggressive collecting program. Teams are sent on expeditions to explore and collect seeds in Africa, Latin America, and Asia. A large seed exchange program with foreign countries has been established and a large seed bank at Kubran has been developed.

The Izmir Center

The Izmir Center was started in Turkey in the mid-1960s. The seeds of the area were sampled extensively and the center proved to be very important for cereals—particularly wheat—and for forage legumes such as alfalfa. Within ten years more than 13,000 seed collections from Turkey itself and at least 2,500 collections from other countries had been assembled. The variation found in Turkey is impressive, and nearly each collected sample is viable. More than two thousand distinctly different lines of wheat were isolated from fifteen samples. The collections here are vital, particularly because much of the native materials are being replaced by uniform crops. Originally the Izmir Center was set up to coordinate activities in a number of Middle and Near Eastern countries such as Iran, Iraq, Afghanistan, and Syria. Political instability and recent warfare in that region may affect the long-term usefulness of the center.

The International Rice Research Institute

The International Rice Research Institute (IRRI) was started in 1960 to collect rice varieties for breeding. The IRRI is one of the most successful examples of a gene bank, exceeding the expectations of its planners. The collection grew rapidly in 1961, jumping from 256 to 6,900 accessions. Rice variants developed from these resources are responsible for the great green revolution and an accompanying increase in productivity. The IRRI now has more than 60,000 accessions and handles requests for tens of thousands of seed samples. Duplications of about 10,000 seeds are kept at the NSSL in Colorado. Today, scientists at the IRRI are attempting to breed resistance to insects and disease into the new rice lines, as well as the ability to grow in drier climates or brackish water. Despite the huge number of accessions already located at the center, some experts fear that valuable wild rice strains that would have contributed much to the new breeding work already have been lost.

The IRRI must rejuvenate its seed every ten to fifteen years. Subfreezing storage would reduce this problem but officials at the institute point out that refrigeration equipment does not last long in the humid tropics. Another problem is that Los Banos is susceptible to earthquakes and typhoons. However, the region is ideal for year-round rice growing.
round rice growing, but is not a sensible place to maintain a longterm cryogenic gene reservoir.

Work at Kew

The Royal Botanic Gardens at Kew, England, has taken the lead in conservation matters for almost a decade. Botanists there have alerted the world's gardens and worked closely with the IUCN on behalf of endangered plants. The first serious gene bank for wild species, as opposed to the existing banks for agricultural species, was started by plant scientists at Kew. Like most botanic gardens, the Royal Botanic Gardens published an annual seed list for some

4,000 to 5,000 species. Leftover seed eventually was thrown away and fresh seed collected for the following season. Scientists knew that if the seed could remain viable indefinitely, this would save time, money, and trouble. Since other scientists had already worked out the appropriate conditions to prolong vegetable seed, it seemed logical to try these methods on wild species. A seed unit was set up, first at the Jodrell Laboratory at Kew and later at the gardens at Wakehurst. In tests with more than two hundred wild species, they found that most species accepted cryogenic storage. One surprising bonus was that orchids, primulas, and delphiniums—all of which have small, short-lived seeds—seemed very amenable to the treatment and could be maintained for long periods. The botanists at the Royal Gardens realized, even in this early period, that their cryogenic seed bank could become an important hedge against extinction. The plan was to hold seeds in the bank until species could be repopulated in the wild, but this approach has begun to look impractical. Still, the gene bank concept is valid and the work at Kew has grown and become further refined.

The seed for the bank at the Royal Botanic Gardens initially came from plants cultivated at Kew. They hoped to obtain seed from other botanic gardens' seed lists and then grow the seeds for several generations until they had enough to store in the bank. However, seeds from cultivated sources tend to have a restricted genetic base, with each new generation containing genes that allow the plants to succeed within the narrow environment of the grower. Only the so-called best flowers are frequently selected. Wide sampling of seeds from wild populations provides a much broader gene pool. The costs involved in breeding several generations for particular seed are very high. It is cheaper to travel to other countries and collect seed directly. A recent study using 1981 monetary values showed that it costs $49 (not including travel and collection costs) to bank a sample that has been collected in the field, while a sample that involves regeneration costs about $260 per sample. The study pointed out that regenerated seed could be obtained from only 44 percent of the species brought in from other gardens.

The gene bank at Kew is the model for other gardens, but unfortunately just a mere handful of gene banks for wild species have followed the one at Kew. We can only hope that more gardens will create their own banks during the next decade.

The Bank in Irvine

At the University of California Irvine, (UCI), we read about the Kew seed bank in 1976 and believed the idea was sensible. But we had no wish to go out and collect all the world's flora for posterity. We felt it was more reasonable to develop a rather small, deliberately defined collection. Outside of Kew, it seemed generally agreed upon that seed banks should focus on the native flora of a particular region. This seemed laudatory and necessary, but we also sensed that duplicate collections should be set up—primarily as a hedge against catastrophe, but also because we believe that while species may occur in certain geographical areas, they do not belong to political nations. Plant species belong to the entire planet. A number of gardens and foundations were already promoting native California species; so we decided to collect South African bulbous and cormous plants. Two families were selected for particular emphasis: the Iridaceae, the iris family, and Liliaceae, the lilies. The iris family includes the genus *Gladiolus,* an important multi-million dollar cut flower crop. We decided to specialize in gladiolus but not neglect other species in the two families. By focusing on defined groups, we know what species we do not have and thus can make an effort to fill in the gaps in the collection. This is more efficient than a shotgun approach in which we would merely collect species at random. The gene bank at UCI has several components—a living, growing plant collection, and cryogenic seed and pollen banks.

We set up the UCI gene bank not only to help preserve endangered species, but also to demonstrate that a small garden can set up a cryogenic gene bank which is economical and does not require technical expertise.

Other Wild Banks

A few gene banks that are not strictly agricultural are beginning to pop up in temperate areas. Gene banks to preserve endangered and regional plant species have been initiated by the University of Dublin, the Polish Academy of Sciences, and the Polytech University in Madrid.

A cooperative effort among countries in the Mediterranean region is aimed at saving at least part of what is left of their wild species. The Iberian Gene Bank in Madrid has been expanded to include all of the Mediterranean flora, including European, North African, and Asian countries bordering the Mediterranean. This effort is called the Artemis Project. Botanists collect seed in the wild and send it to the gene bank. The seeds to be saved are from species considered to be "narrow endemics," that is plants which occur in one small area and nowhere else. The seed will initially be cryogenically stored at the bank at the Polytech University. At some future time, half the sample is supposed to be returned to the country of origin—after a network of banks throughout the Mediterranean region has been established.

Botanists who work on the Mediterranean flora—much richer than the rest of Europe—are organized into a group called OPTIMA (Organization for the Phyto-Taxonomic Investigation of the Mediterranean Area). Most Northern European countries have only tens of endemic species, but the Mediterranean countries each have endemics numbering in the hundreds. Two functions of OPTIMA are to list threatened plant species and to establish a network of seed banks. The Artemis Project is the first step.

Reluctant Gene Banks

As more and more plant species become extinct in the wild, directors of botanic gardens will find themselves the owners of important collections of endangered species. They will be the unwitting custodians of rare plants—sometimes the last remnants of a particular type. An international network of botanic gardens recently was set up to track the scattered remnants. The Botanic Gardens Conservation Coordinating Body started in 1979 as an outgrowth of the IUCN's Threatened Plants Committee and had nearly one hundred member institutions one year later. One of its first efforts was to circulate among member institutions lists of threatened species and to ask which of the species the gardens had in cultivation. So far lists have been developed for European and South African species and also for cycads and succulents. Based on the survey, we now know that 34 percent of the 1,878 threatened species of Europe and 37 percent of the 1,042 South African threatened plants are cultivated in botanic gardens. These studies point out which species are commonly grown and which ones will require special attention. The lists also alert gardens as to which plants among the ones they grow are actually endangered. Learning that a species is endangered can be quite a surprise since the plants may be commonly cultivated. At the UCI Arboretum, we grow lots of South African bulbous plants. When we scanned the lists covering these species, we found many plants listed that we had thought were still common in their natural habitats. One in particular, *Gladiolus guenzii,* grows so easily for us that we routinely used it as a "guinea pig" for experiments, thus throwing away the seedlings after use. But in South Africa, where this unusual species grows on sand dunes, the natural populations have been decimated. We were shocked to find it on the endangered list.

Some people may be pleasantly surprised when they learn that certain plants in their collection are endangered and more valuable than originally suspected. However, the realization of just how many species fall into that category is anything but pleasant. Botanic gardens that possess extensive collections of endangered species bear a special responsibility that accompanies the care of these plants. Unfortunately this responsibility translates into increased attention and eventually into increased costs.

Although some botanic gardens will be reluctant guardians, they will, in fact, carry the mantle of responsibility for many endangered species and must realize that one of their functions is the conservation of endangered plant species. This conservation function has been discussed extensively, but for some botanic gardens this lip service is the extent of their efforts.

Garden Varieties

Besides the need to save wild species, garden varieties of plants need to be conserved as well. As older varieties fall from favor, new varieties of chrysanthemums, carnations, and daffodils are made by the major hobby groups. Over the years, approximately 17,000 to 18,000 new varieties of daffodils have been made. There are more than 60,000 named cultivars of iris. Not all of these varieties need to be preserved, but certainly some should. In October of 1978, the Royal Horticultural Society in England convened a conference on the role of gardens in the conservation of rare and threatened plants. Discussion focused on the disappearance of garden varieties as well as the decline of wild species. Five categories of important cultivars were designated. The first category encompassed historically important hybrids that reflect fashion during a particular period. The second category was for genetically important plants that might carry genes useful for breeding desired traits such as late season flowering or dwarfness. The third was plants with unusual characteristics, such as the doubleness trait in *Hippeastrum* hybrids that have not received serious attention. The other two categories contained species threatened in the wild and species from isolated areas where access is limited. Resolutions were drawn up to confront the particular needs of garden plants, and the participants discussed the real problems to be encountered by their program.

Plant societies were approached to develop lists of important cultivars; then a series of recommendations was set up. Certain plant groups had already established collections in gardens. For instance, old-fashioned roses are a common adjunct to rose collections. Collections of several woody groups, such as Hypericums and Hydrangeas, exist at the National Trust Gardens in England. Although we won't be able to evaluate the success or failure of this project for ten to twenty years, this may be a model for other countries to follow.

Saving Clones

Not all crop plants can be stored as seed. Many fruit trees and bushes are selected clones that would not come true from seed. The only method we have to preserve these clones is with collections of mature trees. Officials at the United States Department of Agriculture, concerned that these varieties could be lost, have created the National Plant Germplasm System—a network of institutions that includes both universities and state or federal agricultural stations. The University of California, Davis facility will maintain a 70-acre orchard of fruits, grapes and nuts. At Oregon State University's Corvallis campus some 1,500 different clones, including over 700 varieties of pears are currently grown. (It's rather amazing how many hundreds of pear varieties have been bred and how few find their way to the market.) A third facility in Geneva, New York, will focus on apples and eastern domestic grapes. We hope they'll also grow wild grape species. Cryogenic storage of tissue cultures will eventually be possible; then we may be able to dispense with the expense of maintaining mature orchards.

What's Next?

Cryogenic gene bank collections are probably the cheapest and most sensible kinds of genetic preserves. In the future, we may find even more economical ways to maintain cryogenic gene banks. Perhaps we could have gene banks associated with either Arctic or Antarctic research stations, where subfreezing temperatures can be maintained without expending energy and brownouts don't threaten collections. Still feasible, but even more futuristic, would be to orbit gene banks in space. There, temperatures could be very low, provided the banks are shielded from the sun's rays and background cosmic radiation. Burial on the moon or on an asteroid might be the best storage system for the earth's genetic heritage. If that scenario seems far out, consider this: within the near future, the analysis of the DNA in a plant's chromosomes will be routine. The sequences of chemicals that make up the chromosomes will be known and stored within a computer. Perhaps the ultimate gene bank will be a computer readout. The genes and chromosomes could then be resynthesized and inserted into a plant egg which has had its nucleus removed and made to develop the desired species. Far fetched? No. Just a decade or two away.

Narcissus viridiflorus

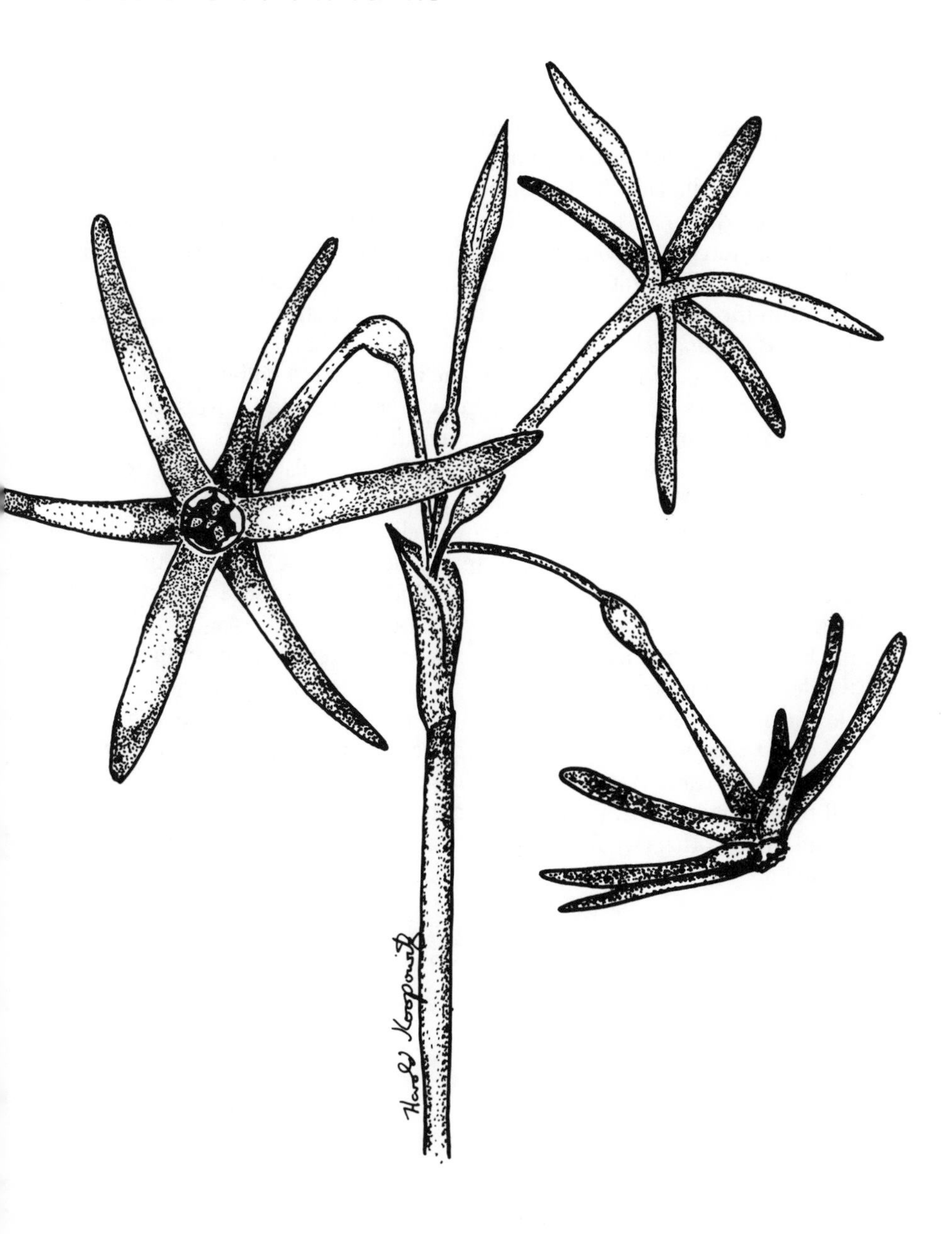

Case History: The Green Daffodil

A youth lay down in the forest clearing and gazed at his own reflection in the pond. Entranced with his own golden beauty, he pined away and died. The gods replaced the lad with the flower Narcissus. This may be just a myth, but we can see that man today is still so enamored with his own worth that he ignores or forgets the world around him. Unseeing and uncaring, he allows the planet to deteriorate—even if this means his own doom.

Within the beautiful narcissus group is a rather strange-appearing daffodil, *Narcissus viridiflorus*. This is a grass-green flower that is almost indistinguishable from a stalk of grass. Each year the underground bulb produces a single stem that serves as the leaf. The stem will produce one to five starry flowers in favorable years. Except for their unmistakeable narcissus fragrance, they can hardly be compared to their relatives, the yellow daffodils of spring. Unlike most daffodils, *N. viridiflorus* flowers in the fall, not the spring.

Many botanists believe the green daffodil is primitive—a remnant of the ancestral forms that gave us our spring flowers. The cup or trumpet that characterizes all daffodils is tiny in *N. viridiflorus,* reduced to just six little ridges. Native to the warm Mediterranean coastline and islands, this species has been steadily reduced. Pressure comes from overcollecting bulbs and from the development of the region. Once relatively common, these bulbs now are almost unobtainable.

At first this species seems to be merely a curiosity with relatively little horticultural value. Yet its genes could permit the development of a race of fall-blooming daffodils. Purists may contend that daffodils should be confined to the spring; others would gladly welcome a breath of spring just before winter closes in.

IV.

THE ROLE OF THE HOBBYIST

CHAPTER 14

Plant Societies

Plant hobby groups are a double-edged sword. These groups may be the best hope for saving certain species because they have a vested interest in those plants. But, these same groups may contain a small number of unscrupulous people whose avarice for rare plants frequently works against the best interests of the species. In this chapter we will briefly survey some of the ways plant hobby groups can become involved in conservation. The following two chapters will examine orchids and succulents in more depth—two groups of plants with wide appeal to hobbyists.

Truthfully, among the various plant hobby groups, very few are directly affected by plant extinction rates—at least for now. Zinnias, chrysanthemums, and springtime daffodils are man-made hybrids whose wild relatives hardly compete with their super-endowed garden cousins. Some plant hobbyists do grow wild species for their own sakes and tend to band into organizations to promote their favorite plants. Their associations encompass a wide range: small, specialized groups that favor peperomias; larger ones dedicated to gesneriads and begonias; huge organizations that promote cacti and other succulents; and the big daddy of them all, the American Orchid Society devoted to the advancement of orchid culture. These societies differ in size and scope, but they have at least two attributes in common—recognition of the value of wild plants, and some sensitivity to the fact that their avocation may have led to the dangerously low numbers of some species.

Rock Garden and Alpine Plants

Generally these plants are mostly wild species growing at higher altitudes near mountain tops. The plants tend to be tufted, making a dwarf cushion of vegetation, and frequently bear flowers disproportionately large for their size. Best-known examples are the indigo gentians and a vast array of stonecrops and saxifrages. A number of alpine societies are spread around the world and some have active seed banks and seed exchanges. These societies have rescued at least one species—*Tecophilia cyanocrocus,* a delightful royal blue crocus-like flower from Chile that is now extinct in the wild. The species has been kept alive in cultivation by the alpine enthusiasts. On the other hand, many problems faced by wild cyclamen species can be laid directly at the door of the rock garden growers. Plant providers have been ruthless about importing endangered cyclamen species.

Wildflowers

Enthusiasts of wildflowers fall into a variety of groups. Many states and countries have hobby groups devoted to the protection, conservation, and cultivation of their wildflowers; frequently these groups do an excellent job of educating their members. Hobbyists learn about the natural history of the species that occur in their region, as well as problems concerned with conservation and protection of wild materials. Wildflower societies have promoted the development of preserves for individual species and frequently present their causes in the political arena. In some enlightened countries, hobbyists are permitted to enter areas to be developed and rescue wild species before bulldozers clear the land. Unfortunately, relatively few countries have such organizations, and even those that exist may not be very effective. Only a very brave or sometimes foolhardy person in a Third World country would stand in the way of "progress." A plea to preserve natural resources, when demands for those resources are overwhelming, is never well-received.

Gesneriads

These comprise a large group of plants ranging from the familiar African Violets, to the recently popular streptocarpus, to a slew of little-known and very difficult tropical genera and species. Gesneriads illustrate a little-recognized fact: when a species is either very difficult to grow or very easy to propagate asexually, only those plants that grow well may be widely cultivated; the rest of the species invariably are lost. Good examples of this among the gesneriads are *Sinningia coccinea* from Brazil and *Titanotrichum oldhamia* from Taiwan. In *S. coccinea,* we see little variation—all cultivated seedlings are almost identical. All of the *T. oldhamia,* a medium-sized plant with bright yellow, waxy flowers that are brown on the inside, are derived from a single clone.

Ferns

The American Fern Society, probably one of the few groups devoted to lower plants, has an extensive spore bank. Few people recognize that as tropical forests are cut down, we lose not only trees and flowering plants, but also a vast constellation of mosses, ferns, and other primitive plants. These lower species need as much protection as the more popular higher plants. The American Fern Society's spore bank is set up for the convenience of its members as a source of rare species. While members recognize that many fern species are on the verge of extinction, the spore bank is not seen as a hedge against extinction but rather as a source for rare species.

Bromeliads

There is an active hobby group concerned with these plants in the pineapple family. These rather curious members of the Bromeliaceae frequently are found as house plants. Most species have leaves that wrap around each other to form a vase. In nature, rainwater collects inside the vase and specialized cells at the base of each leaf absorb water and nutrients. The few species with roots use them merely for gripping and holding plants in place, usually onto a tree branch, a dead post, or even telephone wires. One genus, *Tillandsia,* has members that have adapted to desert conditions. Many have such fantastic shapes that they look almost extraterrestrial, drawn from a science fiction nightmare. The popular *Tillandsias,* often called "air plants," frequently are purchased as curiosities. Usually attached to a twig or a piece of bark, the silvery gray color of *Tillandsias* makes it difficult to tell if they are dead or alive. Most of these *Tillandsias* slowly die in captivity.

Tillandsias available on the west coast of America are mostly from drier parts of Mexico, where entire areas have been denuded of plants. Plant people who regularly visit Mexico describe how they must travel further and further afield each year to find new sources of these plants. Most bromeliads grown by hobbyists are tropical and are rapidly disappearing with the rain forests in the New World. Since many of the more colorful plants grown by hobbyists now are hybrids, we may fiind that many species are lost before their worth has been realized.

Begonias

Nearly everyone has heard of begonias, even if they cannot visualize these plants. This is one plant name recognized even by non-gardeners. What many plant people do not realize is the fantastic variation that occurs within the single genus *Begonia*. Wild species are found growing in South America, Africa, and Asia, as well as in other scattered localities. Hundreds of species have been recognized. As might be expected with such a well-known genus, a very active begonia society is devoted to those plants. The American Begonia Society has a special seed bank that carries many of its more unusual species.

Many other plant-oriented societies exist—some are old and well-established; others are young, growing, and vigorous; and a few are old and senescent. All of these groups can and should play an active role in conservation. Any plant society that currently maintains a seed bank for its members could easily build a cryogenic bank. Since nearly all societies are concerned about the fate of the wild species that form the foundation of their hobby, it is logical for them to take the next step. For these groups, the machinery is already in place. The techniques are easy, and the commitment is there; they just need to be prompted into action.

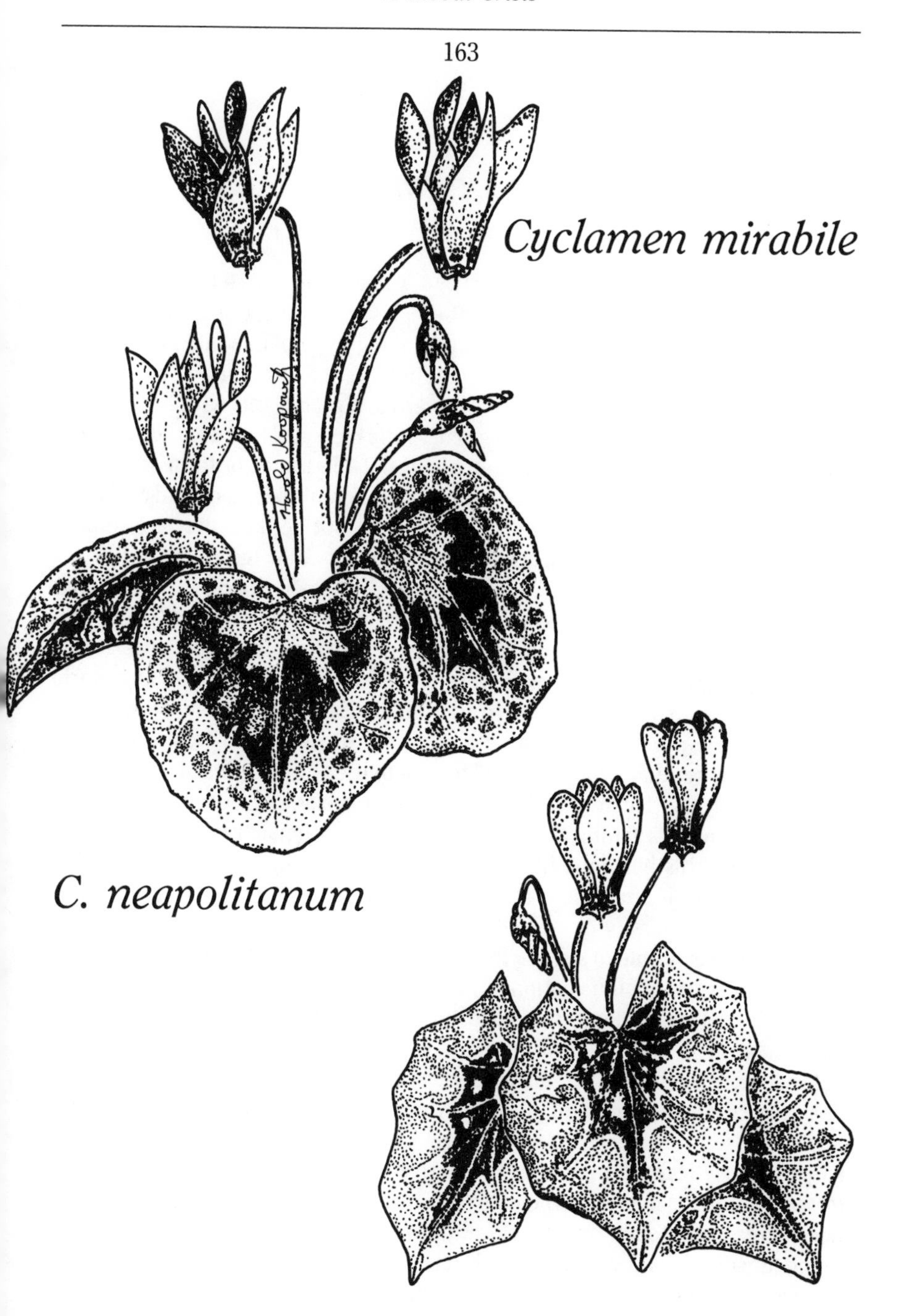
Cyclamen mirabile
C. neapolitanum

Case History: The "Sow Bread" Cyclamen

This is a sad tale of mistaken identity—a tale of how greed and ignorance inadvertently led one already-rare plant species to the brink of extinction. The story concerns several species of cyclamen, a family of plants nicknamed "sow bread" because in medieval times their tubers were fed to pigs. The cyclamen species we see today in florist shops are overfed, potted plants with heavy, waxy flowers. In contrast, the wild cyclamen species are diminutive and delicate, popular as rock garden plants. Hundreds, if not thousands of these plants are sold each year to gardeners. More than twenty different species of wild cyclamen are scattered through the Mediterranean region, but because they are so popular, most of these delicate species have been over-collected in the wild.

Perhaps the most popular of all the wild cyclamen species is *Cyclamen neapolitanum,* also called *C. hederacea.* This plant sports pink or white shooting star flowers and clusters of marbled leaves which resemble tidy, little stacks of ivy. Thousands of these plants were offered to the gardeners of the world for decades, primarily from sites in the Middle East and Turkey. Most of the collectors were peasants who gathered a little extra income by collecting and selling these popular wild plants. Their activities helped sustain their own existence but also led to the depletion of wild cyclamen. Today, trade in all wild species of this family is prohibited by international treaty. Even so, we shall see that this has done little to protect these plants.

Despite these international laws, an English firm placed a large order for wild collected tubers of *C. neapolitanum* from Turkey. *C. neapolitanum* is a hardy plant, easily able to survive the rigorous English winters. This transaction took place despite the international treaty banning such trade. When the 10,000 tubers arrived in England, the importers noticed that the new plants looked different from the *C. neapolitanums* sold to English gardeners on previous occasions. These new plants had a corky tuber rather than the smooth, beet-like surface of the earlier plants. Confused about the shipment, the English importers sent a few specimens to Kew Gardens for identification. To their surprise, they learned that what had arrived from Turkey was not *C. neapolitanum* but instead an extremely rare species called *Cyclamen mirabile.* These new plants, reported in only two localities in Turkey, were anything but hardy. Being unsuited for cultivation in English gardens, the *C. mirabile* did not survive the harsh winter. An entire species was thrown into jeopardy in one fell swoop. In those years, *C. mirabile* was considered rare. Now it is believed to be nearly extinct. The international treaty was designed to prevent such tragedies. In this case it failed to work.

Several questions remain unanswered. Why were wild *Cyclamen neapolitanum* collected in the first place, since there is such a plentiful source of cultivated material for this species? Second, why did the Turks allow the protected plants to be exported? And, finally, why did the British insist on importing plants that were known to be protected? If either country had decided to honor the treaty, the near extinction of this delicate, wild species would not have occurred.

Cyclamen species may still be common as garden plants, but this does not mean that the wild population should be neglected. The days when the species were so plentiful they could be used as pig fodder are long gone.

CHAPTER 15

Orchids: Myths and Realities

The most intriguing mythology within the plant world undoubtedly centers around the orchid. For many people, orchids are synonymous with huge, floppy lavender or white flowers worn proudly next to the heart as a corsage and reserved for those extra-special occasions. Part of the mystique may result from the perception of orchids as very rare, very expensive jungle plants. Popular mythology tells us that orchids need to be grown in hothouses that are a cross between a steambath and a swamp. The myth continues with the assumption that only the rich can afford to grow orchids since the plants thrive only on an expensive, elaborate culture system and that the exotic plants themselves are costly. Orchids, then, are status symbols. Those who cultivate orchids obviously are at the peak of their avocation or profession.

Like most myths, a grain of truth is buried beneath the tall tales. Here the kernel of truth is very, very small. Many orchids boast flamboyant, gorgeous flowers, but relatively few of these plants are exceptionally expensive and most are not that difficult to grow. In fact, many orchids can easily be grown in the home.

The Precarious Life

The orchid family is the largest group of flowering plants, consisting of more than 7 percent of the total number of plant species. Conservative estimates suggest at least 20,000 different orchid species in existence. New species are still being discovered. Orchids are not confined to the tropics but actually range across the continents. Species grow at the tip of South America in Tierra del Fuego as well as in Alaska; but in the tropics of the Americas and Asia, orchids really flourish. While just 211 species have been named in all of the continental United States, more than 2,600 species occur in Papua, New Guinea alone.

Not all orchids have large flowers. Some species are very tiny and insignificant, yet even these small versions turn out to be quite interesting when their biology is understood. Many orchids have flowers that are designed to be pollinated in strange and wondrous ways. For some species, the reality is even stranger than the fiction that surrounds the entire group. For example, *Coryanthes,* the bucket orchid, secretes so much nectar that insects that slip into the pouch-like petal that stores the fluid must swim out to avoid drowning. In doing so, the insect is guided past the plant's sexual structures, thereby pollinating the flower. Another group of orchids, *Ophrys,* have petals that mimic female insects. Male insects are fooled into copulating with the flower and, while doing so, pollinate the plant. These are just two of the many stranger-than-fiction examples found in the orchid kingdom.

Orchids depend heavily on the insects that pollinate them. Therefore, ecological changes that affect insects and pollination can spell disaster for the plant species. Other disruptions to the ecology have deleterious effects, too. Orchid seeds are tiny balls of undifferentiated cells that leave the mother plant lacking the food supply contained in most other kinds of seeds. Seed can live in the wild for only a few weeks. During this short period, the seed must arrive in the correct nook or cranny where it will be able to grow. Some orchid species like to perch on the branches of tropical trees, while others spread their roots in soil or through the crevices of rocks. The seed must find the exact habitat that is appropriate for that species. This life-or-death matter happens strictly by chance. Once the location has been found, the seed must be invaded by a particular species of fungus. Together, the orchid and the fungus set up a mutually beneficial relationship. Of all the thousands of seed produced, only a small percentage survive. Charles Darwin once illustrated this point with his calculation that if all the seed produced by a single orchid plant were to survive, the plants from those seeds would cover an entire acre. And, if all the seed produced by those plants were to survive, their offspring would pave the British Isles. The fourth generation of plants would cover the entire land surface of the earth! The fact that we are not constantly stepping on orchids points out the high mortality rate of the seed.

The few plants which find the correct niche, and the correct fungus, require many years to mature to flowering size. During the first few years, the plant is particularly vulnerable since it is small enough to be killed by a small grazing insect or snail. If the plant survives all of these obstacles, the probability of the correct pollinator finding and pollinating the bloom is only about 5 percent during any season. In casual surveys we have made in the wild, we have learned that, on the average, just one in twenty flowers actually ends up with mature seedpods. No wonder orchids frequently are rare. In order to permit the best possible seed production, orchids have evolved ways of "tricking" pollinators to carry all of the pollen produced by the flower in a single packet. Pollination then becomes an all-or-nothing event. Either the maximum number of seed or no seed at all will be produced.

Orchids and their Habitats

Orchids occupy two main habitats: epiphytes cling to other plants (usually trees) and geophytes grow on land. Most orchids are epiphytes and live in the forest canopy. Ten to twenty different species growing on a single tree is not uncommon and frequently trees may be loaded down with even more species. However, a collector may have to search many square miles to find another tree that supports the exact same species. Many other popular plants—such as bromeliads, philodendrons and ferns—are epiphytes and they share the same habitat with orchids. A few orchid species are very successful and highly competitive and, if they settle on a tree, will occupy it to the exclusion of all other species.

Epiphytic orchids are the ones most commonly grown by hobbyists. Methods have been developed for growing their seed without the participation of the fungus. This has led to the mass production of hybrids and also many of the species. Some species are very rare and endangered in the wild but are quite common in cultivation. The beautiful *Cattleya skinneri* is an example of an orchid nearly extinct in its native Costa Rica but which is readily available to the hobbyist because it can be grown from seed. More plants of this orchid currently exist in North American greenhouses than probably ever existed in Central America.

The geophytes present a different picture. Because these plants are more closely tied to their symbiotic fungi, consistent germination has eluded most plant scientists. A group in Australia, however, has learned to germinate their geophytes by culturing the fungus and adding it to the seed culture. Unfortunately many of these terrestrial orchids are not easy to maintain as adult plants, either. Mature plants seem to hate being transplanted. Among the desirable but difficult-to-germinate terrestrial orchids are the *Cypripediums,* the temperate cousins of the tropical *Paphiopedilums.* Populations of these orchids growing in Great Britain have been monitored carefully since before 1900. The naturalists have watched the numbers dwindle until just a single plant remained. The overall prognosis is not good. Some of the African terrestrials also are being grown from seed now but it is still a hit-and-miss affair, with most species continuing to die.

The Role of Collectors

Two distinctly different dangers threaten the existence of wild orchid species. One danger is over-collection by hobbyists and professional collectors. The second is deforestation, in which numerous orchid species are destroyed when the forests are cut down. The vast majority of orchids are epiphytes and are most threatened by the loggers who make their way through the forest. Collectors also go after prize specimens in this group. Geophyte orchids are less numerous but even more threatened, since these land-loving orchids are almost impossible to grow from seed.

You can still find garden magazine advertisements selling North American terrestrial orchid species such as the fringed *Habenarias* and various lady slipper orchids. Some of these plants may flower the first season but they usually deteriorate. Just a minute percentage survive more than two or three years. While the number of terrestrial species offered for sale has dropped in recent years, this does not necessarily reflect enlightened conservation attitudes. More likely it is an indication that the supply is growing scarce. Do editors of gardening magazines still run these kinds of ads out of ignorance or are they merely indifferent to the fate of yet another species? Or is the dollar most powerful again?

Orchid hobbyists are among the most enlightened groups of plant people. Most hobbyists have been aware of the conservation crisis that faces wild orchid species for many years. A quiet controversy has rumbled through their ranks as the hobbyists mulled over a central question: what share of the blame must hobbyists bear for the deteriorating species? Certain species obviously owe their rarity to their desirability and responsibility falls to one group. For example, *Odontoglossum crispum* is a crystalline white orchid that captured the attention of European orchid hobbyists during the last century. The cool, moist, mountain forests of Colombia were ravaged for the *Odontoglossums,* which were sent by ship across the Atlantic. Many plants perished enroute. The best survivors commanded high prices and this led to additional searches for more plants. Whole forests were cleared to get to the orchids perched in the trees. Cutting down a tree for the orchids was easier than climbing the tree trunk to tear off the plant. The natives realized that the wild population of these flowers was approaching extinction. By then it was too late. At one time they contemplated introducing *O. crispum* to the mountains of Jamaica as a hedge against extinction, but nothing ever came of the plan. Whether or not the cool-growing *O. crispum* could have survived in Jamaica was a moot point. Perhaps it was just as well, since the introduction of a foreign species often upsets the natural ecological balance of the habitat.

Many of the prime *Cattleya* and *Paphiopedilum* (paphs) orchids have also been collected out. Most of the orchids imported and/or smuggled into the United States are "paphs," such as *Paphiopedilum purpuratum,* a small, dark purple lady slipper orchid that was localized in the area of Hong Kong. At one time, *P. purpuratum* was so plentiful and common that it was used as an aquarium plant. Tropical fish fanciers had discovered that the orchid continued to live inside fish tanks. The plant gradually drowned. Four to six months later it would have to be replaced. This short lifespan did not matter in those days, since the species was very common and very cheap. Eventually the plant became collected out and advertisements began to appear in the United States exhorting hobbyists to buy up the remaining plants while they were still available. The political tension between Hong Kong and the People's Republic of China came into play. A three-mile barrier—essentially a "no man's land"—stood between the New Territories of Hong Kong and the Chinese mainland. When conditions grew less tense, the barrier was reduced to a single mile-wide stretch of land. At that point, collectors rushed in, found an unsuspected patch of *P. purpuratum* and collected it out. Nobody knows whether more of the species exist in the remaining narrow strip or if the species has truly been lost. This was one surprising instance where political strife actually provided a sanctuary for a species while peace, in contrast, resulted in disaster.

Orchid hobbyists have realized the advantage of adding genes from wild species into their modern, complex hybrids. Through the twentieth century, despite the popular steady line breeding, hybridizers have continually gone back to the wild species to use their genes to enrich the breeding lines. *Cattleya aurantiaca* is a Central American orchid with small and not particularly pretty flowers. This species does have two outstanding features: the flowers are an intense, pure orange; and the petals are thick and heavy. The petals are so thick that they look like plastic. Using this species, hybridizers were able to enrich the *Cattleya* group with a variety of heavy, yellow-orange and red flowers. The shape features were provided by the other parent, usually a standard hybrid. In a similar vein, *Cattleya luteola* is being used to make the currently popular miniature cattleyas, and *C. bicolor* is contributing green shades. Who knows which species will contribute tomorrow's fashionable flowers? More importantly, will the desirable species still exist when they are needed? Many other orchid species are being collected out as discussed in the case histories scattered through this book.

Many more orchids are endangered by factors other than the disastrous effects of collectors. A number of orchidologists have pointed out this discrepancy to soften the blame they feel has been directed at them. Some point out that hobbyists who go out in the wild to collect their own plants do little damage and hardly ever travel inland more than a mile from a roadway. Experienced hobbyists usually do collect in a responsible fashion, but we have seen many enthusiastic beginners strip a tree of all its plants—many of which are too fragile to stand the shock of travel or transplantation. Professional collectors are even more troublesome. Hobby collectors have reported returning to areas once rich in plants only to find that the areas have been stripped by professional collectors.

Deforestation

Deforestation most seriously threatens the tree-dwelling epiphytes. The number of species exported from the tropics is minuscule when compared to the number destroyed by logging operations. It is difficult to estimate how many plants are killed with each tree that falls, since some tropical forest trees carry several hundred plants and others are bare. Many geophyte orchids also are destroyed in logging by the machines that trample the ground. Those individuals that survive are often fried by the tropical sun since their shade has been removed.

Many different people, familiar with different parts of the tropics, tell us the same story time and time again. They tell of visiting a particular hillside covered with verdant forest, an area rich in epiphytes and an incredible assortment of other plants. Within a year or two the people return to the same area and find nothing but bare earth. The forest has gone. With it has gone thousands of ferns and orchids.

Can Anything Be Done?

The future looks grim for the many orchids that live on trees directly in the logger's path. Can anything be done to save these species and the genetic diversity within them?

One suggestion stems from the fact that people grow plants because they enjoy them. Even in slums you can find a lone plant flourishing in a rusty tin can. People have suggested that growing orchids could be a cottage industry—that is, peasants in underdeveloped countries could grow different orchid species and sell them cooperatively. Something like this was set up in Mexico, but in recent years little has been heard of the experiment. Too many orchid species are threatened in too many parts of the world for this type of project to be effective. Orchid farming might be a profitable cottage industry but it would not take long for an astute person to realize that growing orchid hybrids would be more profitable than growing orchid species. Furthermore, many species might have to be wild collected which would devastate whatever plants still exist in the wild.

In the short run, preserves can be helpful. These parcels of protected, undeveloped land have limitations, but we need all the preserves we can get, both large and small, besides living orchid species collections. Perhaps most importantly, we must save orchid species in cryogenic gene banks. We need collectors out in the field, collecting as many seeds as possible from different parts of a species' habitat. This way, hundreds of thousands of different individuals can be saved. Operation of a seed bank does not require the extensive knowledge of species ecology necessary to grow live collections. The ideal situation would be for individual countries to set up their own gene banks, but that is unlikely to happen. Instead, the American Orchid Society could set up and maintain orchid gene banks, just as other interested groups could maintain banks for their particular group of plants. So far, few societies have stepped forward to begin such projects.

Orchid collecting is a necessary evil. We could support a ban on orchid collecting if it meant that species would be protected in the wild. This usually is not the case. According to the Convention on International Treaties in Endangered Species (CITES), all orchid importations are forbidden except under special permit. This was designed as blanket protection for all orchid species, endangered or not. Unfortunately, it has retarded the rescue of orchid plants from areas undergoing active deforestation. Mounds of plants that could have been saved are destroyed and allowed to rot because of red tape and bureaucratic snafu. Restricting trade in endangered species is essential, but we do no service by imposing restrictions that prohibit preservation and protection of vulnerable species. We see no evidence that the CITES laws are effective in the case of orchid plants. Shipments are still coming into the country, plants are still collected in the wild, and the forests are still being cut down.

Peristeria elata

Case History: The Holy Ghost Orchid

Being elevated to the level of the national flower of a country can be devastating, particularly if the plant also assumes a supernatural or religious aura. This tale is about one of these rather unlucky species.

Peristeria elata is a curious plant. Each year it grows stems, or pseudobulbs, that look like green, upside-down turnips, perhaps a little larger than the average turnip. From the top of each stem, a few long and wide leaves emerge, arching up and out. Each leaf is folded lengthwise. A cluster of pseudobulbs and leaves create a luxurious tropical plant, but it is not this lush vegetation that has caused near-extinction. The real attraction is the flowers. A thin, erect flower stalk grows four to five feet tall out of each mature pseudobulb and carries one or two rows of small, globular flowers. From a distance the flowers scarcely seem remarkable. But as you approach the plant, you notice their unusual feature. Each small blossom consists of five waxy petals. Inside the petals there is the unmistakable figure of a white dove, formed by the fifth petal and the plant's sexual parts. The natives believe the flower represents the Spirit of the Trinity in a marble sepulcher. Most people, therefore, refer to the plant as "The Holy Ghost Orchid."

The plant's fame spread far beyond its native Panama. *P. elata* became a collector's item. Although relatively common and easy to obtain in Europe and North America, back in Panama the Holy Ghost Orchid is struggling to exist. The combination of exportations for hobbyists, demands for church flowers, and habitat destruction have placed *Peristeria elata* on Appendix I of CITES. Despite the endangered status and the fact that cultivated plants are available at low cost in North America, the remaining wild plants are still being collected.

We remember opening a shipment of Panamanian orchids confiscated by customs officials because documentation was missing. The plants had been handed over to the university for safekeeping. The individual plants had been neatly labeled, including even the more obscure species. Except one. There, at the bottom of the box, lying *incognito*, was an unlabeled *Peristeria elata*. The unscrupulous collector surely knew the species' status but was too greedy to resist. There's no way to tell how many times this scenario is repeated with endangered species.

CHAPTER 16

Cacti and Other Succulents

The terms cactus and succulent are often confused. Although people sometimes use these words interchangeably, they do not mean the same thing. While all cacti are succulents, not all succulents are cacti. Succulents are plants with swollen leaves or stems that store water for the plant in case of drought. Many plant families have members that are succulents, including the Cactaceae plants, commonly called cactus. Another incorrect notion is that all cactus plants have spines. The truth is, some cactus plants have no spines at all and some plants from other families do have spines. Now that those misunderstandings are straightened out, let's move on to see how these plants are faring in the battle for survival.

Cactus plants and many other succulents are not doing well. Figures taken from a 1975 report by the Smithsonian Institute show that 84 percent of the plant species considered to be exploited for commercial gain were succulents. Of these, half were cacti. Unlike orchids, where most trade involves man-made hybrids, trade in cactus and other succulents usually involves species. A collection held by any succulent enthusiast is largely composed of pure species since wild plants already have such incredible variety and shape.

Popular Plants

By their very nature, succulents are difficult to kill. Adapted to harsh, dry desert conditions, succulents can go for many months without water. They die slowly, waiting for a change in conditions that will permit their recovery. This makes them ideal house plants, especially for people with less-than-green thumbs. Unlike ferns which will wither and brown if not cared for, a succulent will still be there in six months, even if it is neglected. Since many succulents can be propagated from seed or reproduced from cuttings, they usually are relatively inexpensive. Provided they are given enough light, succulents will grow in the same conditions that are comfortable for humans. Many species are easy to grow and easy on the pocketbook, although there are some "connoisseur" species that are difficult to obtain or grow. Some very large specimens can be expensive.

The easiest way to obtain a good-sized succulent is to buy a wild-collected plant. This is the main reason for the plant's problems. Some cacti are notorious slow-growers, taking ten years or more to reach a respectable size. Because humans tend to want instant gratification, digging up a healthy, mature specimen has been considered the best route to take, both for the consumer and for the nurseryman. By not having to raise succulents from seed or cuttings, the nurseryman avoids the costs of labor, space, and supplies. International laws have somewhat slowed down the wild-collecting. But, as we shall see in some of the ensuing examples the laws came too late for some species and in many cases are terribly ineffective.

The natural rarity of many succulents compounds the problem. Some plants were thought to be extinct, but decades later they would resurface again in a new region. Sometimes a plant mistakenly acquires a reputation for being rare because the people giving the label are not familiar enough with the country or area of origin. More often, a plant that is common in cultivation may be on the threshold of extinction in the wild. Since some plants are so well known in gardens, they do not receive the protection they require in the wild. The *Aloe variegata,* commonly called the Partridge Aloe, is a very common potted plant in England and Europe and is also relatively easy to obtain in the United States. However, this plant has been collected out in its native South African habitat.

About twenty years ago, while hiking on the peak of a low hill in southern Africa, Koopowitz spotted a small clump of a succulent with silvery leaves. It appeared to be a sister species of the rosary vine, a popular houseplant of the milkweed family. A piece of the plant was sent off and identified as a very rare plant that was considered to be lost. Fifteen years later there was no sign of this rare plant at the same site. We do not know if it finally succumbed to natural conditions or to man's efforts. This species may no longer exist in the wild, but it has been grown quite easily in California for the past twenty to thirty years, even though most people don't know how rare it really is.

Another succulent milkweed thought to be extinct both in the wild and in cultivation is called *Whitesloanea crassa,* named after Alain White and Boyd Sloan, two succulent enthusiasts who produced three important volumes on the Stapeliads, a tribe of the milkweed family. Many of the Stapeliads are popular succulents that have rather weird starfish-shaped flowers that smell like rotting meat or fish. *Whitesloanea crassa,* which resembles a weathered cube of rock, was rediscovered in 1957 and brought into cultivation. Despite further searching, no more plants were found in the wild. Plants lingered for a while in various gardens, but all of them died in the end. It's possible (though unlikely) that *Whitesloanea* still exists in the wild. Perhaps a few plants escaped the Somali goats and are still clinging to stoney ridges in their native habitat.

The greed of collectors has threatened *Pediocactus papyracanthus,* whose name means "foot cactus with paper-like thorns." This species' major locality is in eastern Arizona, but few remain there now. A single collector went into the region and dug up every specimen, large or small, that he could find. His actions alone jeopardized the entire species. While Arizona does have strict conservation laws, the state is large and the conservation enforcement section is small. Secondary roads which are filled with cactus cannot be patrolled; so the piracy continues.

The *Ariocarpus fissuratus,* or living rock has been plagued by collectors, but we still do not know just how threatened this species has become. This is a strange-looking plant that can best be described as a cross between a buffalo chip and a handful of coarse gravel. There are no spines on this cactus. The living rock cactus grows sunken into the earth with its almost-flat, fissured surface level with the ground. Its gray-tan skin blends into the earth for camouflage and protection. The plant looks dull as dirt for most of the year until the correct season when it produces lovely rose-pink blossoms. This species has been extremely common but also is very slow-growing and equally slow-dying. Only ten years ago you could go into a drug store in California and buy a box of Texas Living Rock for just $1.50. Most of the plants sold in those stores were more than 25 years old. We know that the stores no longer sell these strange plants, but it's difficult to tell just how many are still alive in the wild.

Big Business

With the exception of just one genus, all of the cacti species are New World plants. Many species are found in the United States, where about 25 percent are considered rare. Among the rare species are the very large cacti such as the big, red barrel cactus of Southern California and the famous saguaros from the southwestern deserts. These species take a very long time to reach an impressive size and command high prices. These cacti are protected in states such as Arizona, but are big business on the black market. Illegal saguaros are sold by the foot; in Europe or Japan, a saguaro cactus goes for at least $40 per foot. How these huge plants—sometimes six or seven feet high—are smuggled halfway around the world may involve some rather intriguing enterprises. It would make an interesting story. Illegal cactus dealing is a million-dollar business in Arizona alone. A small band of law enforcement officers in Arizona attempt to thwart this business, but they estimate that a successful cactus rustler can walk away with $100,000 per year. Cacti can be collected legally in Arizona but the process involves money, time, red tape, and obligatory inspections. For the unethical who are not disturbed by breaking the law, cactus rustling is much, much simpler. Even so, more than 1,000 permits are issued each year in Arizona for people to collect species legally out of the desert. Those familiar with the southwestern deserts claim that they can follow the decline of the desert vegetation. Deserts may appear to be indestructible, but they are actually fragile and easily converted to lifeless zones.

Areas in which a particular cactus species is dominant, such as the saguaros of Sonora, depend upon the species as an important factor in the region's food chain. Half of the bird species in Sonora depend directly on the saguaro cacti for food.

The Population Crunch

Desert plants are being plagued by more than the collectors. The saguaros, for example, are being uprooted from their natural desert homes and moved to developing suburban housing tracts for landscaping. Theoretically, these tall cacti should be able to survive and prosper among the new homes, but we have some doubts. California once had large groups of saguaro cactus but few of these plants exist today. The ones that have been transplanted have succumbed to old age and various ailments without setting a new generation to take their place. We fear that this may also be the end result with the Arizona saguaros.

Developers have also disturbed the genus *Bulbine*, succulents in the lily family. Some of the *Bulbines* look like *Aloes*, with thick, triangular leaves; others have cylindrical, pencil-shaped photosynthetic organs. *Bulbines* have yellow, star-shaped flowers, sometimes clustered on long stems. A species of *Bulbine* with bright orange flowers used to grow near Cape Town in a small area that was suddenly bulldozed to make way for an apartment complex. Some of the plants found their way to England and the United States. Although obliterated in the wild, the species has been saved.

Succulents are threatened not only when land is completely cleared for new housing but also when agricultural needs claim undeveloped land. For example, *Echinocerus reichenbachii* var. *albertii,* one of the Texas Hedgehog cacti, hides in dense thickets of spiney mesquite. This species will find itself in greater jeopardy as machines clear more and more land of mesquite.

With any alteration of the environment, some plant species are bound to be adversely affected. A peculiar cactus called *Pediocactus knowltonia* ran into trouble when the San Juan River in New Mexico was dammed. During the dry season (which lasts most of the year), the species retracts its stem and is completely covered with dust. The plant, scarcely an inch in diameter, hides in this position until the rains finally come. The plant absorbs rain water, swells, and finally emerges above ground. The pink flowers it produces are dramatic—nearly twice the size of the cactus' body. Hiding underground helped the plant until the river was dammed, when most of the species' habitat became flooded. How ironic that after the species spent thousands of years evolving its physiology to survive drought conditons, it may ultimately be done in by drowning. According to the Red Data Book, this species is not expected to survive beyond the mid-1980s. Only two populations, each with a few hundred individuals, are left. Heavy recreational activity and the rising water level of the dam are threatening these remnants.

Disappearing Lone Star Cacti

Texas seems to be the worst state for the future of cactus species. A large number of desirable cactus species exist in Texas, but the combination of range extension, brush clearance, and over-collection is threatening large numbers of plants. According to eyewitness accounts, huge warehouses and barns are filled with cacti waiting to be shipped out of state. Inside the Big Bend National Park many species are protected, but not all of the valuable species are located inside the park. One of the finest collections of small, spherical cactus is located outside the park. Unlike the gigantic saguaro and barrel cacti, the spherical cacti in Texas are just right for window sill potted plants. In the mid-1970s there were reports that these cacti were being removed at a rate of about a half-million plants per year. Collectors were selling them at ridiculously low prices—just $18 per thousand to whomever came to cart them off. Vast areas were stripped of their cacti. We suspect that these plants were potted up individually and retailed at fifty cents to one dollar each. The original collectors did not get rich, but the middle-men must have made a fortune. One eyewitness claims to have seen a shed with twenty piles of assorted cacti. Each pile contained at least a thousand plants. Not only did collectors scoop up plants outside of the national park, but they also poached inside the park, going after *Epithelantha bokei,* a particularly desirable and now threatened species. Park rangers discovered huge piles of cacti—all dead. The poachers had dug up the cacti and then, for some unknown reason, left them behind to rot.

The push to bring marginal lands into cultivation in Texas and other southwestern states is a potentially serious problem. Acres of natural arid vegetation are being sprayed with herbicides and cleared by bulldozers. The surface is then dressed with fertilizer and pasture seed. The biggest problem with this is that no one really knows if there will be sufficient rainfall to maintain these lands as cattle pastures. Sometimes rare and valuable cactus species may be wiped out for no good reason.

South of the Border

Ironically, while cactus populations in the United States are being decimated by illegal collectors, large quantities of cactus located south of the border are being destroyed needlessly. Plants are being destroyed in areas of the Mexican desert that are being converted into farmland. Vegetation mechanically is pushed to the edges of the fields where it is left to die and decompose. An ideal solution would be to allow these plants to be collected and sold to gardeners in other parts of the world. This would help fill the worldwide market for cacti without stripping areas of the southwest that are not being bulldozed for housing or farming. Most people are dissuaded from legal collecting by the red tape involved in obtaining an export-import permit from the Mexican government.

While many succulents in the United States and Mexico are facing disaster, species in other parts of the world are also in trouble. A few years ago an illegal shipment of young Euphorbias from Madagascar was confiscated at the port of Los Angeles. When mature, the plants form cactus-shaped trees, but they are fragile when young and need to be grown in a heated conservatory. The average glass house would not be sufficient because there would not be enough room for the plants when they reached maturity. These plants were doomed from the start.

This example points out a problem faced by a number of botanical gardens in the United States. As customs officials become more educated about embargoes on endangered species, they confiscate more plants. Usually the illegal plants are sent to a nearby institution until the importer gets the correct permits. If the importer is not permitted to reclaim the plants, the institution is asked to maintain the endangered plants. This certainly is better than permitting the illegal importing to continue, but too often the institution is saddled with a collection of rare plants it does not want or need. Sometimes, the expense of maintaining an endangered plant with very specific needs is more than some institutions can bear.

The market for cacti seems to be insatiable. England, Germany, Holland, and Japan are the primary markets. Illegal collectors are tempted to circumvent the law because the wild-collected cacti are so easy to sell. Overseas collectors sometimes visit the natural New World habitats and do the collecting themselves—often with disastrous results. We've heard of an instance when a Japanese collecting expedition traveled to Cedros Island off Baja California and stripped the island of all its native succulent plants. Incidents of botanical piracy must be stopped if succulents are going to survive at all.

While illegal collecting has not abated, we have seen an increase in the propagation of many kinds of succulents, including cacti. Visitors to South Africa can see farms where large numbers of American cacti are being grown quite rapidly for the European market. And in California, huge wholesale nurseries are propagating South African succulents by seed and cuttings for sale in the American marketplace. Through improved methods, substantial profits are being made by those gardeners.

Hobbyists Can Help

Hobbyists dedicated to cacti and other succulents are among the most common plant collectors. They have banded together in a number of different societies—all dedicated to enjoying and furthering these plants. What role can these societies play in the future survival of succulents growing in the arid zones? The potential power of these groups is tremendous, but it's difficult to say whether they actually will wield the power they have. The problems faced by succulents have been obvious to hobby groups for a number of years. As a result, many cacti societies have committees devoted to conservation of rare species. How much good these groups have done so far is difficult to determine since more and more succulents are threatened, rather than less and less.

Societies can help in a number of ways. One simple way is to conscientiously educate their members about the need to conserve and protect all species. While the society as a group may be very aware of the problem, some members may be totally unaware that the problem is a serious one that should be considered by anyone involved with plants.

A second way to help would be to request that all members purchase only legitimate plants, thereby helping put the illegal collector out of business. Many people come across plants that are obviously illegitimate but reason that since the plants already have been collected and can't be returned to the wild, they might as well enjoy them. If everyone started to shun these illegal plants, collectors might begin to get the message. Several years ago one of the most prestigious nurseries in Pasadena, California proudly advertised "wild-collected" clumps of *Lithops*—a rare stone mimic from southern Africa. The plants may have been obtained legally, but even if they were, their loss in the wild puts a needless strain on the wild population. *Lithops* are easily grown from seed so there was no need to sell or buy wild-collected specimens. Feedback from local succulent enthusiasts might have persuaded the nursery not to participate in similar sorts of ventures in the future.

A former member of a conservation committee from the largest succulent society pointed out the problems faced in changing people's opinions and actions. He couldn't help but be pessimistic about the committee's sincerity regarding conservation since he felt that given the chance, even the committee members would buy a very rare plant regardless of how it was obtained.

Societies could help preserve plant species by changing some of the ground rules of plant shows. Currently, the largest specimens win the awards. For slow-growing species, this means plants that are very old. Smaller and younger individuals of the species would not win. The result of this standard is that for the most part only wild-collected individuals win awards—a sure incentive to continue the wild-collecting. If the policy was changed so that only nursery-propagated stock could be winners, the demand for wild stock certainly would diminish. This is a very concrete contribution that societies could make to the overall problem.

Assessing the Future

In December of 1981 about fifty people—hobbyists, government officials, scientists, and nurserymen—met in Tucson, Arizona, to discuss the effects of cactus trade and collecting on the remaining wild populations of cacti. Several interesting facts came to light during that meeting. First, it was reported that in 1979 about seven million cacti and other succulents were legally admitted into the United States from other countries. About one and one-quarter million cactus plants came from Mexico where they were collected in the wild. After a short stay in the United States, the plants were exported to other countries with the claim that the plants had been propagated in U.S. nurseries. This ploy—representing a wild plant as a nursery-propagated plant—is also seen frequently with orchids.

Those attending the meeting repeatedly pointed out that while protective laws exist at many levels—state, federal, and international—the real problem is enforcing those laws. Manpower simply does not exist to enforce adequately the rules already on the books. The near future does not look any better. Even when the laws are enforced—as they often are in California and Arizona—penalties are relatively minor. People engaged in illegal trade can still make good-sized profits even after paying the penalties if they are caught. If rich countries such as the United States cannot control this illegal trade, poorer countries such as Mexico will be even less able to control the problem. Although all Mexican species of cactus are included on Appendix I of the Convention on International Treaties in Endangered Species (CITES), the country has just one cactus nursery. That means that nearly all of the exported cactus plants are wild-collected.

We can see that laws designed to protect these endangered plants probably will have little effect on the overall problem. Looking into the near future, one ray of hope may be the rapid replication of endangered species by using tissue culture techniques. This may help save some species on the brink of extinction, but as far as preserving the variation in the species' gene pool, this cloning method will not help. Still, some variation is better than no variation. These tissue-cultured plants may fill the needs of the hobbyist and relieve pressures on wild populations. One of the first cactus species that may be saved with this method is *Pediocactus,* the cactus in Arizona that was nearly wiped out by a single collector. The succulent *Aloe polyphylla* has already been tissue cultured and distributed by Kew Gardens. Technology must be perfected to the point where nursery stock can be grown more profitably than obtaining plants from the wild.

Finally, cryogenic gene banks can be furthered by the various succulent hobby groups. Since many groups already have a seed bank or seed list, the extra effort needed to put together a cryogenic gene bank should not be a problem.

Aloe polyphylla

Case History: The Spiral Aloes

Lesotho, a tiny land-locked country in Africa, is so mountainous that sometimes it is referred to as the Tibet of Africa. Its winters are bitterly cold with snow and temperatures often well below freezing. Usually less than forty inches of rainfall occurs in the summer. In this environment we find the *Aloe polyphylla,* more often called the spiral aloe.

Aloes typically are desert-dwelling plants. Being succulents, these members of the lily family usually are well-adapted for hostile, arid environments. Their thick, triangular leaves store fluids to carry them through long periods of drought. The spiral aloe has the most unusual environment of all the aloes and, in fact, dislikes dry, hot weather. Its home in Lesotho is at an altitude of about 8,000 feet, in the seepage areas on the western mountainous slopes where there is abundant moisture.

Spiral aloes can be very beautiful, forming a large rosette of triangular gray-green leaves arranged in five spiral series. Some plants display a clockwise spiral, others counter-clockwise. Transparent ridges line the margins of each leaf and some may even have a few teeth. In the springtime the rosette throws out a candelabrum of coral-colored flowers.

The geometry of the clean, spiral arrangements of the leaves makes the plants so desirable, even though mature plants can become as large as two feet in diameter. Since spiral aloes are avidly sought by collectors of succulents, natural populations have been decimated. Estimates of the surviving numbers range from 200 to 3,000 individuals in the wild.

The Lesotho government prohibits export of both seed and plants and currently is trying to establish a nursery for commercial purposes. Even so, the prognosis for this species is not good. The plants are difficult to maintain outside their natural environment and are most particular about their water requirements. Because the plants look as if they come from a dry, sandy desert rather than from a moist mountain top where they are bathed directly by clouds, collectors are confused about how to care for spiral aloes. Most succulent hobbyists do not realize that the plants prefer cool temperatures and abundant moisture. Even though spiral aloes are treasured by collectors, the plants usually are doomed once they are collected.

Seed of *Aloe polyphylla* was smuggled out of Lesotho and seedlings eventually found their way into the trade in Southern California. Some of these plants are now approaching maturity. We hope that when they flower, seed can be produced and this desirable species propagated for future generations to enjoy.

CHAPTER 17

The Ethics of Collecting

The International Organization for Succulent Plant Study (IOS), based in the United Kingdom, has generated a code of conduct for its members to follow. The code is excellent—one that is pertinent to all plant groups, not just those that collect succulents. The code is aimed not only at hobbyists but also at scientists, professional plant collectors, and others interested in wild species. In the following pages, we summarize and discuss the main points of the code and add some elaboration of our own.

Conduct in the Field

When you collect wild plants, the first rule is to disturb the population as little as possible. Try to leave mature seed-bearing plants; if that isn't possible, just take cuttings or seed. You may believe that a species is common and feel no compunction about collecting an entire hillside's plants, but subtle differences do occur between plants of the same species located in adjacent areas. Even if it is more trouble, gather some plants from different populations and localities. This will insure that the subtle variations are not lost.

Mature specimens frequently do not transplant as well as younger individuals. The time of the year that you collect can be critical to the plant's welfare. You often need to know your plants well to determine which season is best. For example, many plants in the iris family will not transplant when they are in flower. Numerous bulbous and cormous plants, such as those of the *Narcissus* genus, make storage organs only after flowering. For many years, wild *Narcissus* species were collected by Spanish peasants in the spring during the active growing season, often when in flower. The bulbs were allowed to dry and were then shipped to Holland. In the fall, the bulbs were distributed to the nursery trade and sold. Few plants survived. The fact that a handful of bulbs actually sent up a flower or two is testament to their remarkable fortitude. Most plants merely threw up a leaf and expired. The best time to dig is usually when plants are becoming dormant.

You should also assess whether or not it is even possible to grow the plant out of its native habitat. Species that live in very specialized environments—such as acid bogs or cool, moist forests—should be left to experts. Some of the most exquisite flowers are almost impossible to grow.

Investigate local ordinances concerning plant collecting. Conditions vary considerably, not only from country to country, but also from state to state or county to county. If a permit is required, obtain one and follow the restrictions. The permit may define the species and indicate the number of plants that can be taken. If a report is called for, be accurate. Do not collect officially designated or even unofficially considered endangered species or plants. One special exemption is when plants are taken into cultivation for a propagation reserve. A private individual should not undertake this task. If you collect specifically to propagate an endangered species, do this in cooperation with a local institution—either a botanic garden or university. You may want to try to tissue culture for endangered species but, if you do, avoid irrevocably mutilating the parent plant while you obtain the needed tissue. You cannot tissue culture all species—some are quite recalcitrant. Many plants are self-sterile, such as the succulent aloes, and cannot be sexually reproduced unless two different individuals are used.

If the plants to be collected grow on private land, do not trespass. Get permission first, preferably in writing, from the farmer or landowner. The IOS suggests that those who wish to collect plants may be more successful if they notify the local plant society, if one exists, and make themselves known to local plant enthusiasts. This is not only a polite gesture but plant people usually are cooperative. Locals who know the flora can be of immense help in locating desirable plants. A word of warning: don't be greedy.

Collect moderately and do not give local residents any ideas about the commercial value of their plants. If you are collecting orchids and you find a very rare variant such as an albinistic form that could give the plant impressive value, act nonchalant. If the locals have the impression that the entire population is worth a fortune, the plants will not last long—even if the valuable plant was the only good clone for miles around.

Some native plant collectors make a living by finding plants and selling them. Such people frequently have secret and favorite collecting areas and may be prepared to collect for you but not take you along. Make sure that you provide the collector with adequate descriptions, preferably with drawings or photographs, so that he or she does not strip a dozen species hoping to collect the one that you want. Pay a fair price. Paying five dollars a plant you know may be worth several hundred dollars reeks of avarice. On the other hand, overwhelming generosity can generate wholesale collection of a species that only you had wanted. The plants then would be left to die or rot for lack of buyers.

Don't cut down a tree, for a rare fern or orchid may be growing out of reach in the tree's canopy. Usually a large variety of epiphytes perch on a single tropical tree. We hope that people will no longer leave trails of destruction and devastation through forests.

Collect notes and data about the plants. When you return home you may not be able to remember which plant grew in the sun or on the western side of the hill and which one came from the side of a stream or a dry, arid patch at the base of a mountain. Meticulous data about the plant's habitat and its growth in the wild will be useful for growing the plant successfully. We have seen forest floor plants that like to hide in the deep shade succumb after being bleached by direct sun in the wrong part of the greenhouse.

The IOS also encourages people to try to assess the vulnerability of the wild population. With repeated trips to an area, anyone can soon tell if the population is diminishing. Make a survey of possibly endangered species and estimate the number of mature and immature plants. This will help evaluate how rapidly the population is dwindling and give your report more credibility. If a construction project is to go up in the near future—perhaps a road or a dam—the collector can alert other interested parties and save those endangered plants.

Before packaging and bringing materials home, make sure everything is cured, cleaned, and properly packaged. Wash the soil off the roots and remove all dead leaves and stems. A toothbrush is good for scrubbing plants clean. Do not bring in plants that are obviously diseased. Carefully go through the plants, looking for insect pests which can hide very effectively. Investigate all nooks and crannies. Small round holes bored into stems are always suspicious. Cut into them and make sure they are empty.

Cacti and other succulents need to be packed dry. Newspaper makes a good absorbent but don't package plants so tightly that you make a compost heap. Remember, if one plant starts to rot it will spread to the other plants. It's a good idea to cut off very large leaves or trim them back to one-quarter of their original size. If you use plastic bags, blow up the bag and then tie it off to give the plants breathing space. Don't fill the bag with everything that can fit into it. You aren't making a salad. Label all plants clearly and concisely and cooperate with customs and agricultural officials. If they determine that your plants need to be treated for insects, request spraying rather than fumigating. Dry plants survive fumigation better than wet, another reason to have them relatively dry during transit.

The Nurseryman's Code

The world is filled with unscrupulous people trying to make a fast buck. The plant world unfortunately has its fair share. Some nurserymen are happy to import the last wild-collected individual of a species, seeing only their profit. Perhaps these nurserymen rationalize that they may as well trade in these plants because if they don't, someone else will. Most of the specialty nurseries are run by honest people who love plants and are fortunate enough to combine their work and their pleasure. They unknowingly can cause trouble. If they import or deal in wild plants, these well-meaning folks can play a role in the devastation of species.

The code for nurserymen designed by the IOS stresses propagation of species. Nurserymen are asked not to sell or even advertise unpropagated wild species. To this we add that they should advertise that the stock they offer is, in fact, nursery propagated. A number of orchid nurseries propagate their own species stock and mention this in their advertising. There is the growing realization among many plant importers that times are indeed changing. Supplies of wild-collected plants are diminishing and the more discerning nurserymen foresee the time when imports will play an insignificant role in plant trade. For the present, we are in the critical period when people are anxious to obtain the last few collections of certain plants. Propagated plants that mature slowly are more costly than wild-collected materials. It can be cheaper to grow some plants for sale from asexual cuttings but this tends to lose genetic diversity. The IOS encourages nurserymen to use several different clones for propagation. While it might make good business to keep exclusive control of rare species, it is better for the survival of the species that plants be distributed. This is insurance against disease and occasional catastrophies.

Nurseries need to be careful about labeling. By printing on the label the collecting localities or the areas of origin, nurserymen can add a dimension to plant hobby collections, and possibly increase sales.

Changing Attitudes

Hobbyists need to change their attitudes dramatically. People need to realize that they do not "possess" a species. Instead, hobbyists are the stewards of the species—guardians with certain responsibilities. The main responsibility is the survival of the species. As more and more species become extinct in the wild, more private individuals will find the last survivor in their collection. A new moral responsibility will need to be recognized as the private sector plays an increased role in the custodianship of plant species. The fact that individuals bear these responsibilities is rarely discussed. The IOS suggests that the important criterion for a good hobbyist collection is not the rarity of the species in the collection, but the way in which the plants were grown. Good cultivation should be the goal.

Growers should not patronize suppliers who sell unpropagated wild material. Buy plants bearing a label stating that the source is nursery-propagated stock. Hobbyists should not hesitate to inquire about the plant's origins and refuse to buy those that are not homegrown. Only in this way can you let collectors know that wild materials uprooted for quick profits are not acceptable. This is one way to slow down the ravaging of wild populations. Plants collected and rescued from construction sites are exceptions to this rule.

If you believe that plants offered to you come from wild-collected stock and were obtained illegally, report this to the authorities. Also, bring the matter to the attention of editors of garden journals where these firms or collectors advertise.

Hobbyists should be careful about their own labels. In the future, the labels of collected plants may be the only source of identifying lost species and their localities. Be careful how you copy the label and make new ones. In this age of plastics, plant labels have a short, fragile life before they become brittle and fall apart. Replace the label long before that time.

If you grow rare material, try to set seed and propagate the plants. If the first attempt is unsuccessful, try again and again. Sometimes plants will not set seed until conditions are just right. Try to find another hobbyist with the same species and cross pollinate. Share your seedlings—spread them around. The best investment against loss of your own plants is your generosity. Countless hobbyists have related how lucky they were to have given cuttings to friends and been able to replace a prized specimen after a mishap.

* * *

Nearly all plant societies have flower shows where hobbyists exhibit their prized specimens. As we pointed out in the chapter regarding succulents, judges at flower shows should give more credit to well-cultivated material than to rare, collected material. Judges should also give preference to well-grown but immature species, rather than mature but obviously imported plants. Such changes would help reduce the pressure to collect wild plants.

These subtle changes in direction need to be adopted and publicized in show schedules, meetings, and at judges' refresher courses. We are somewhat pessimistic that this will actually occur on a large scale. Most of these points developed by the IOS have yet to be adopted by the majority of succulent societies.

Iris winogradowii

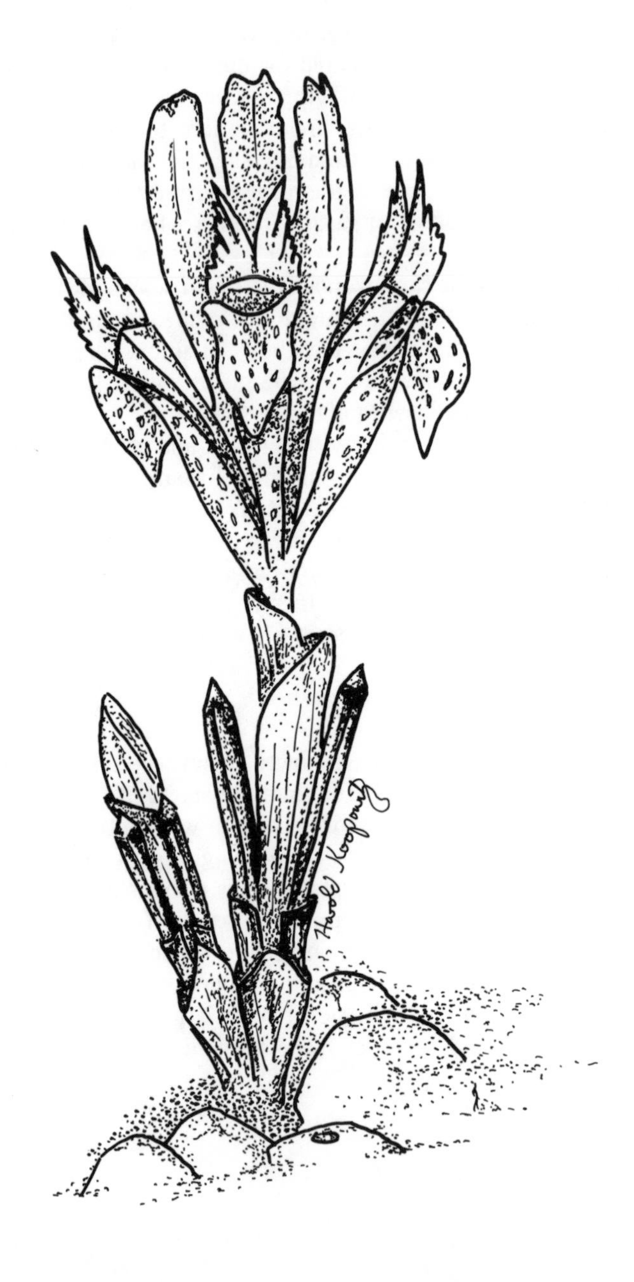

Case History: The Yellow Alpine Iris

Deep in the Caucasus Mountains, in an isolated meadow, a golden yellow iris captures the early morning spring sun and jubilantly hurls it back into the sky. This sunny species is *Iris winogradowii,* a plant that has been known for only the last fifty years. It is confined to sub-alpine meadows on the side of just one mountain near Bakuriani in the Georgia Soviet Socialist Republic.

Most of the species within the iris genus have spectacular flowers. Over the years, many different kinds of garden irises have been bred. This Georgian iris is a relative newcomer and is a rather unusual member of what is known as a reticulata—hardy, early spring flowers that are usually blue or purple. Just two species of the reticulatas are yellow. Rock garden enthusiasts and home gardeners appreciate the dwarf stature of the reticulata irises. Reticulata irises tend to flower even before their leaves emerge from the winter soil. The leaves themselves are unusual, often square-shaped in cross sections and bear stiff points. The lemon yellow flowers of *I. winogradowii* are about two inches wide and bear an orange blaze across the three largest petals. These markings give the flowers their warm glow.

Not long after *I. winogradowii* was discovered, plant collectors practically decimated the population. Fortunately, some of the plants made their way to Holland where they have been propagated and distributed to gardeners and plant people interested in dwarf alpine irises. The species has a good chance of surviving in cultivation, partly due to its attractive appearance and ability to withstand cold temperatures in northern latitudes.

The Russians recognize that the species is endangered in the wild and in need of protection, but whether or not effective measures have been undertaken to thwart its demise is not known. Many plant species in many parts of the world are officially recognized as needing protection, but few actually receive the help they need.

V.

POLITICS

CHAPTER 18

Legislation

Protection for plant species has been recognized as an important aspect of the overall conservation picture just within the last decade. Before then, most efforts focused on the protection of animals, both mammal and bird species. Only a few lucky plant species were favored and protected in specific countries and states.

Today, many conservationists recognize that animal conservation and plant conservation frequently go hand-in-hand. Slogans such as "Eleven animal species depend on each plant species" are beginning to appear and animal preservation societies are getting involved with plants as well.

Although a species may inhabit a particular country, does it belong to that nation? Perhaps we need to regard species as being the heritage of the planet Earth, to be enjoyed and utilized by all people. While people are perhaps too aware of borders between countries, other organisms do not recognize national boundaries. Many birds routinely fly halfway around the world on their yearly migrations. They are no more concerned with the designations of countries than migrating herds of antelope are aware of national boundaries in East Africa. In many cases the ability to conserve a species depends on international cooperation. Quite recently we have seen an increased readiness to help preserve species from other parts of the world. The developed countries of the western world are primarily responsible for underwriting worldwide conservation efforts. Within the United Nations, the Food and Agriculture Organization (FAO) has paid close attention to the need for increased agricultural lands, and the adverse effects this is having on the environment. The FAO has also been concerned with the maintenance and preservation of genetic resources. The FAO's concerns have generated a great deal of international interest.

The Early Days

In the early 1950s, nature lovers formed the International Union for the Protection of Nature and held yearly meetings on conservation topics and problems. The meetings, which attracted a wide group of enthusiasts, tended to be emotionally charged. Max Nicholson, a participant and chronicler, wrote: "Readiness to pass sweeping and strongly worded resolutions was in inverse ratio to knowledge of the relevant facts as a whole and to the capacity for securing action for them." By 1956, when the name was changed to the current International Union for the Conservation of Nature and Natural Resources (IUCN), the meetings became more rigorous and technically oriented. The new organization received official recognition from the United Nations and from major world governments, all of which sent representatives. The IUCN was then carrying out its own reconnaisance mission and holding fewer meetings. In 1958, the group brought out the first edition of the United Nations World List of National Parks and Equivalent Preserves. Two years later, the IUCN turned its attention to the ecology of large, hoofed herbivores, such as the African ungulates, and organized a conference in Tanganyika on these species. Africa was still reeling from both the euphoria and malaise of the independence movement. IUCN officials were consequently pessimistic about the effectiveness of the conference. Much to their surprise, African nations proved to be very sensitive to the problem. Max Nicholson, in his chronicles, related that the African delegates showed substantial concern and understanding. The African representatives, he said, pointed out that the Europeans had already exterminated their own large herbivores while the African nations managed to keep theirs.

The next conference focused on the wetlands of Europe and North Africa, ecosystems in trouble. Plants and animals adapted to moist environments were seriously threatened when marshes had been drained. By 1965, the IUCN once again turned to the tropics. A conference on the conservation of tropical Southwest Asia was held in Bangkok and triggered considerable activity.

During the 1960s, people began to realize that conferences were wonderful consciousness-raising events; however, there was a big gap between the good-willed intentions of the participants and the actual resources needed to put effective conservation programs into action. The IUCN had formulated scientific policy and suggested governmental actions, but there were other important needs, too.

Money was needed. To obtain sufficient funds, the public needed to be educated about the need for conservation. The IUCN felt education could be supplied via a public relations campaign focused on conservation problems. Furthermore, legal considerations would have to be considered. To satisfy these additional needs, a new organization was created—the World Wildlife Fund—to handle fund-raising, public relations, and legal affairs. Prince Phillip of England and Prince Bernhard of the Netherlands became prominent fundraisers, which created an international flavor within the organization. The World Wildlife Fund began to publish Red Data Books for endangered animals. Attention had not yet been turned to endangered plants, but there was a slowly growing awareness that many plants were also threatened with imminent extinction.

The Threatened Plant Committee

The first Red Data Book on plants was published in 1970. Conservationists began to realize the severity of the situation as they collected data. The book suggested that as many as 10 percent of the world's plant species could be threatened in one way or other. We now know that this was a rosy estimate. In the middle of 1974 the Threatened Plant Committee (TPC) of the IUCN was set up and headed by Professor Heslop-Harrison, then director of the Royal Botanic Gardens at Kew, England. Since then the Royal Botanic Gardens has maintained an important and prominent role in plant conservation.

The TPC had two functions: first, to accurately assess the threatened plant situation around the world; secondly, to formulate actions to enhance plant conservation. To carry out the first task, regional threatened plant committees were set up to identify threatened plants. Some regions did not have a TPC, but appropriate botanists were asked to compile lists in those areas. Most plants were handled geographically, but plants of special interest—palms, tree ferns, cycads, and succulents—were dealt with on a worldwide basis. Conferences were held at Kew in 1975 and 1978 to consider ways to alleviate the problem. Initial reports from the regional TPCs reflected that the situation appeared to be worse than anyone had imagined. For example, in East Germany, nearly 37 percent of the evaluated species were extinct, endangered or threatened. Eighteen percent of the British flora was threatened. Sixteen percent of Danish plants were in trouble. In the continental United States, 2,140 kinds of its 20,000 species of plants were in danger, plus an additional 1,113 species listed from the Hawaiian Islands. Hawaii's figures worked out to an astounding 97 percent of the endemic flora of the islands threatened. Tropical regions are rather poorly understood; consequently, lists for those regions have still not been produced. Reports have trickled in from many other regions.

The next major step by the TPC was to identify which of the threatened species in 1980 were still safe in cultivation. The TPC created a new organization for this purpose, the Botanic Gardens Conservation Coordinating Body. The first task of this new group was to identify which of the plant species on the various lists were in cultivation; then it was necessary to aid in the propagation of threatened material and coordinate exchanges. The new coordinating group ultimately intends to educate the public about rare and endangered plants, provide research materials, and produce source material for hobbyists and others to reduce illegal collections.

The first surveys taken by the Botanic Gardens Conservation Coordinating Body show that by 1980, more than 520 specimens of the 1,878 threatened European species were located in roughly 40 botanic gardens and more than 235 species of Madagascan succulents were located in various gardens. Similar lists of plants from the southern African and Macronesian regions, as well as lists of cycad plants, have been produced. The group hopes to produce a "Green Book" containing the localities of threatened plants in cultivation.

Botanic gardens recognize that grave problems are involved with the propagation of plant species. The genetic structure of cultivated populations can change; catastrophes can wipe out a collection. Therefore, the group believes that the only safe place for a plant species is in its natural ecosystem.

We disagree. While we hope that large numbers of natural ecosystems can be preserved, it is highly unlikely that more than a few will be set aside. In any country, state, or area, literally hundreds of tiny microhabitants provide specialized ecosystems for individual species. Gross changes in the overall environment occur regularly. Even the stability of natural ecosystems is open to question. If present trends continue, the contribution to conservation by botanic gardens will be much more important. We live in a world of diminishing resources, a place where some preservation is better than none. Maintaining species in their natural ecosystems may be the ideal approach, but our world is not ideal. We shouldn't underestimate the importance of captive populations, especially since plants thought to be extinct in the wild are popping up unexpectedly in cultivation. These "rescued" species range from the trees of Easter Island to some of the wild grasses of Great Britain.

We believe that too much attention has been given to genetic erosion—the loss of genetic variability in captive populations. While this is a problem, obviously small, uniform populations are better than no population at all. The total number of varying genes in a species is small when compared to the "invariant" genes that code for that species. The total genetic structure may change in captive populations from those that once existed in the wild, but the main genes for that particular species will remain and survive. The cultivated plants will not be a new species, but rather may possess slightly different traits. We know from firsthand experience that when plants are bred in captivity, surprising variability not seen in the initial collection will surface. At University of California, Irvine, the first seed obtained of *Moraea loubseri,* now extinct in the wild, produced a half-dozen uniform plants. Three generations later we find several variations in the blooming pattern, flower size, and depth of color. The kinds of variation may not be the same as in the wild species, but the species does continue to exist.

Convention on International Trade in Endangered Species (CITES)

Trade in wild species is big business. While trade is carefully monitored for animals, import and export of plants are difficult to assess. In 1977, at least 38 million plants were imported into the United States, many without permits.

Because many of the endangered animals and plants occur naturally in Third World countries and are transferred to the Western industrial countries, the onus of trade regulation falls onto the shoulders of the poorer, developing nations. To help these countries cope with this problem, the IUCN lobbied for international trade controls and brought about the first Convention on International Trade in Endangered Species of Wild Flora and Fauna (CITES) in 1973. The trade agreement was enforced in 1975 after ten nations had ratified the document by passing legislation based on CITES. By 1980, the original list of ten CITES signatories had swelled to 59 as more countries joined the plant conservation effort.

Three lists known as the appendices are at the heart of CITES. These lists mandate the treatment necessary to import or export species placed on the list. Appendix 1 contains critically threatened species or genera. Transport of any specimen from this list requires two permits—one permit to export the plant from any country and another to import it into the recipient country. Specimens can be moved only if removal will not jeopardize the survival of the species.

Appendix 2 lists species, whole genera, and even entire families of plants that are not as seriously troubled as the plants on Appendix 1. Substantial trade of the plants on Appendix 2 could result in the plant's being driven toward extinction. This list also includes plants which are not threatened but which resemble plants that are. For some groups, such as orchids or cacti, only an expert could distinguish the endangered species. Blanket protection is therefore given to several entire families. Only export permits are needed for plants listed on Appendix 2.

Some plants are rare in one country but fairly common in others. Appendix 3 lists these plants and allows countries the option of restricting trade in those species.

Another organization, called TRAFFIC (Trade Record Analysis of Flora and Fauna in Commerce), monitors international trade of species. A few years ago, TRAFFIC conducted an experiment at major international airports of countries which had signed CITES to see how closely the CITES regulations were being followed. The experiment involved carrying a cactus plant through customs. The plant was either declared or displayed prominently at the customs counter. The United States and the Soviet Union were the only countries to confiscate the plant. This was only because the plant did not have the correct phytosanitary or health permit. The fact that all cactus are on Appendix 2 apparently was not the deciding factor. In the United Kingdom, the customs officials obliged by filling out the health certificate. None of the customs officers encountered during the experiment understood or even seemed aware of CITES, even though some of the officials had a copy of the CITES documentation nearby. While CITES generally was ignored in the early years, the situation may be a bit better today. Los Angeles International Airport now has notices posted that warn Americans leaving the country not to import organisms listed in CITES.

Is CITES Effective?

Quite a few countries still have not ratified CITES. And, even CITES signatories may have loopholes in their legislation that permits them to evade the spirit of the convention.

In April of 1979, a West German tour group returned from a cactus collecting expedition in Mexico. When CITES officials requested German customs to search for the contraband plants, some 3,600 plants were confiscated. Under German law, a total plant importation valued at less than DM240 (almost US $100) is exempt, even if the plant species are listed on Appendix 1. Therefore, many of the cactus plants were returned to the collectors. In California we find that imported plants that do violate CITES' regulations are confiscated.

The major function of CITES may turn out not to be restricting trade, but rather the alerting many governments to the international concern over endangered species. We now see awareness in a variety of countries which had previously been indifferent to the fate of their own flora.

The CITES document itself may threaten species. We have no quarrel with either Appendix 1 or 3, but we believe that the restrictions applied to Appendix 2 may be harmful to the longterm survival of endangered species. Reports from Mexico and South America say that plants are being cleared in those areas to make agricultural fields. Those plants are being left to rot at the edges of the fields because of trade restrictions. CITES covers international trade but not protection in the home country. Without home protection, species can still be destroyed. The major threat to all but a handful of orchids is the deforestation of the tropical forests. CITES unintentionally impedes the rescue of many of these plants. Because they can't be exported as a cash crop, due to the blanket protection of CITES Appendix 2, the plants are doomed to die with the trees that are cut down. We acknowledge that a customs officials probably cannot tell the difference between a *Cattleya skinneri* (Appendix 1) and a *Cattleya forbesii* (Appendix 2), but blanket protection is not the answer. In this day of computer technology, perhaps an extended list with illustrations could be generated for Appendix 1 or 2; suspicious plants could then be referred back to knowledgeable authorities. Blanket embargoes of whole families sound good but are really not helpful in the long run.

Preserves

Without a doubt, the ideal situation would be to set aside numerous parcels of countryside where wild species could grow and flourish without interference. To approximate this ideal, many countries have developed a system of national parks. There is a growing awareness in the tropics that these preserves play an important role in conservation. More preserves are being established, some in previously indifferent countries.

Setting aside a chunk of land on a map and calling it a preserve is not enough. These areas must be patrolled, guarded, and managed. As such, public access to the preserve needs to be limited.

Major problems plague preserves, partly because the preserves frequently are set up for animals, not plants. Plants are protected from humans, but there is little protection from the animals. Animal populations can increase rapidly and quickly devastate the plants. Koopowitz remembers a small game preserve set up near a little village in South Africa that was the pride and joy of the locals who loved to point out how well the antelope were growing and reproducing. No one recognized that as the antelope flourished, the vegetation was being progressively destroyed. Even the bitter leaves of the aloes were being cropped.

To be successful, preserves must be managed by people with extensive knowledge of the natural history of the organisms within the preserve. *Orothamnus zeyheri* is a beautiful protea from South Africa with flower clusters that resemble roses. The plants were devastated by people who picked the flowers; eventually the population was reduced to a single hillside. Officials created a preserve of the area and carefully managed the remaining plants. Nevertheless, the population continued to dwindle down to a mere handful. Extinction seemed certain. One day a fire broke out in the preserve and, despite heroic efforts, blazed through the entire preserve, charring everything in its path. Everyone thought they had seen the last of *Orothamnus* but, the following rainy season, hundreds of *Orothamnus* seedlings germinated. The problem was that the seeds needed an occasional burn to stimulate germination. Now that this is known, the preserve is burned routinely. The species survives.

Sometimes well-meaning conservation attempts backfire or produce negative results. Plants of *Teucrium scordium* were discovered on a piece of commonage used by duck hunters in Great Britain. The rare nature of the plants was recognized, and so a fence was erected to protect the plants. The plants were smothered by the surrounding vegetation that was also protected. During the pre-fence days, the duck hunters trampling around had kept the competitors under control. The *Teucrium* could not have survived without their help. Understanding the ecology of a system is the key to a prosperous preserve.

Vanda coerulea

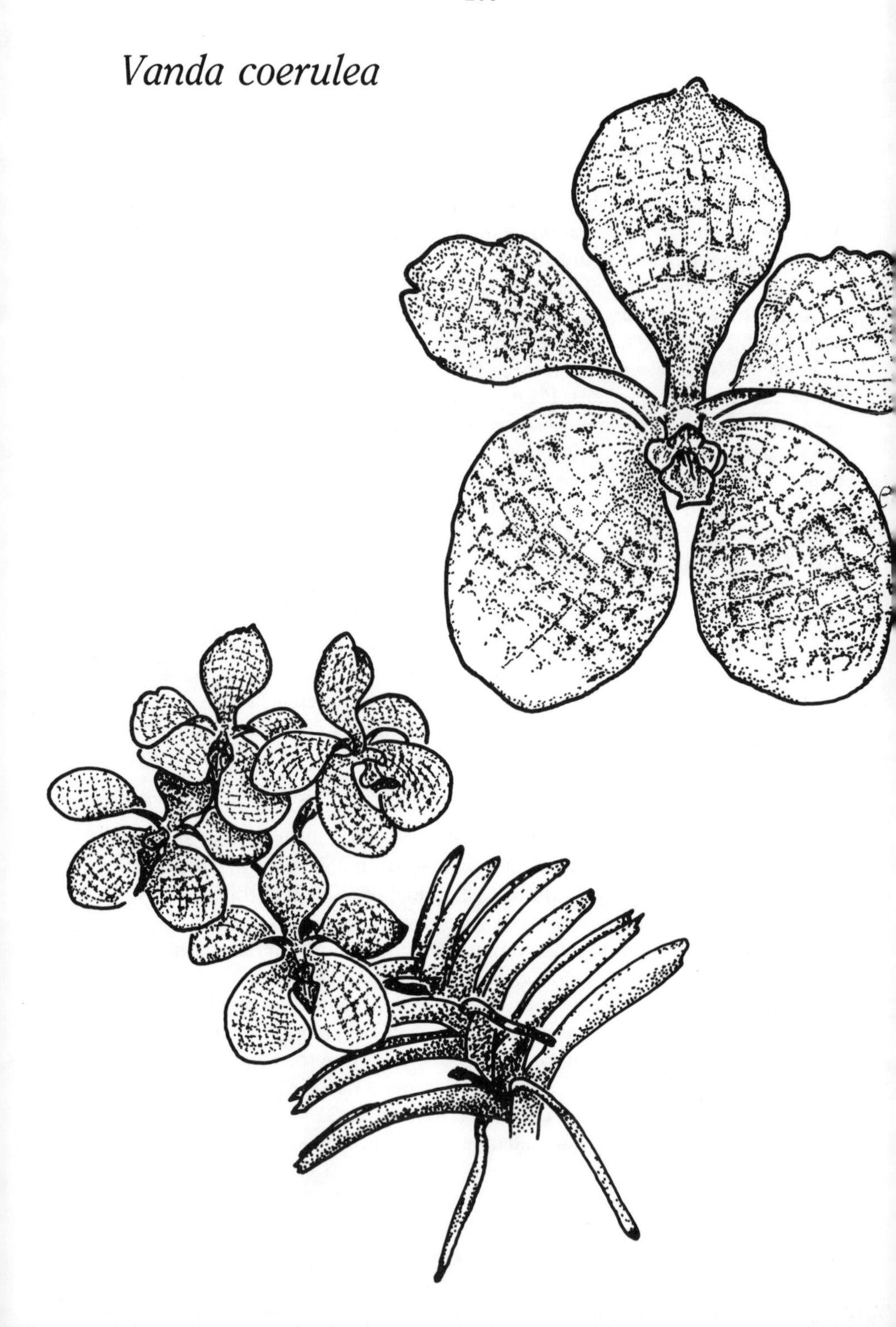

Case History: The Blue Vanda

When the earth broke its back and thrust up the spectacular Himalayas, one of the richest floras in the world was produced. The majestic mountains, deep valleys, and foothills created a myriad of habitats that fostered both vigorous, cold-loving plants and plants that thrive in mild climates. In the Khasia hills of Assam in 1837, Dr. William Griffith discovered a wondrous orchid plant that lived not in the thick tropical forests but rather perched on the branches of small oak trees. Here the plants were exposed to a dry climate and the heavy frosts of winter. The inconspicuous fans of yellow-green leaves held flocks of spectacular blue, moth-like blossoms. Any large sky-blue flowers are rare; large sky-blue orchids are even more unlikely—nearly as rare as guinea pig tails. The flowers ranged in color from clear eggshell blue to a lavender-blue with darker, checkerboard markings. They measured about four inches across and held up to twenty blossoms in a single spray. Named *Vanda coerulea,* the plant generally was known as the Blue Vanda.

The Blue Vanda was not rare in its own habitat. Huge numbers were collected and sent to Europe where they became very popular. Because the plant grew at higher altitudes, it could be cultivated more easily than its relatives that grew at sea level in the dense tropical forests. This easily grown orchid was ideal prey for beginning, amateur orchidophiles. The demand was tremendous. Although the Blue Vanda grew abundantly, very soon there would not be enough flowers to meet the demand. Oak trees were stripped of all their orchids as more and more of these plants were sent overseas. The orchids were unfortunately sent to their new homes via ships and many of them died before reaching their destination. Of those that survived, most were in poor shape after the long journey. These subsequently died soon after reaching England. Plant scientists in those years had not yet devised methods of germinating orchid seeds; there was no way to grow new blue orchids once the imported plants were gone.

Because the blue orchid was so strikingly unusual, or perhaps because the Indian government was just very wise, whatever the reason, *Vanda coerulea* was the first orchid to receive official legal protection from its native government. The protection was designed to help the blue orchid reestablish its normal population levels in the wild. Help came too late. The blue orchid never recovered its normal population and has continued to decline. The survival of this plant is now seriously questioned. Why wasn't it easier to save this unique orchid?

Even with official protection, plant collection and exportation has continued until recently. The Blue Vanda is considered endangered. Greed and personal profit won out over common sense and conservation. Little hope is being held out for the few plants remaining in the wild, but the species is being grown in large numbers in captivity. The Blue Vanda is an epiphytic orchid that can be germinated in the laboratory relatively easily. Even so, what a shame to lose the natural diversity that exists in this species. Orchid growers can use only the limited variety of genes contained in those individuals in captivity.

CHAPTER 19

Getting Involved

In the course of our work we talk, both formally and informally, to many groups and individuals about the plant extinction crisis. Our conversations elicit several different kinds of responses. The first reaction is almost always astonishment. Nearly everyone is shocked by the extent of the problem and surprised that the extinctions have received so little publicity. Some people then assume that the matter is out of their hands because some governmental agency is responsible for solving the problem. Others express outrage and insist on finding out what they personally can do. Those people with a readiness to get involved can accomplish quite a bit. When a problem is big, as is this one, many people must become involved at many different levels. If there are many people involved, each one need only take one or two small actions. We can't save all the species, but it is far better to save a few species rather than none at all. In this chapter we will outline some possible courses of action that lay people without special expertise in plants can take. The list is not complete. You may have even better ideas. In the appendix we offer listings of organizations that you can contact to become part of the solution.

The Political Arena

While legislation is not the total answer to easing the extinction crisis, conservation laws are important. You need to write letters—even short notes—

to local legislators to inform them about your concern about plant extinctions. Let them know that you consider plants to be a valuable part of the country's heritage, and that you would like to see your concerns reflected in their legislative activities. Legislators are confronted with so many issues that you cannot expect them to pay attention to plant extinctions without receiving public pressure. Your letter should be a query requiring an answer from someone on the representative's staff. A simple declaratory letter may only be read and filed. A letter that requests information usually makes a more lasting impression. Perhaps you can ask your legislator:

a) How effective is the current conservation legislation in regard to protection of plant species?

b) What pending legislation might affect plant species' conservation?

c) What are the representative's views on state monetary appropriations earmarked specifically for conservation and plants? Follow each query with a short statement that explains your view of the importance of plant species conservation.

Using the Media

Letters to newspapers should be similar to those written to politicians. Numerous letters are frequently needed before the issue receives attention. Newspapers, magazines, and journals are involved with consciousness-raising among the general public not just the legislators. Not everyone is interested in plants. In fact, many people probably would not be alarmed if three-quarters of the plants became extinct overnight. These people need to be educated to realize why we should care about plant extinctions. They should be taught that seemingly insignificant extinctions can and will affect their lives. Organisms that capture the fancy of most people—whales, condors, elephants, and tigers—tend to be large. Most interest among plants is directed toward large trees such as the giant sequoias (even though one famous politician was quoted in the 1960s as saying about redwoods, "Once you've seen one, you've seen them all!"). Because people do not get excited about small herbs with scientific names, they must be taught why they need to care about obscure plants.

Not too long ago, one of us walked into an office carrying fund-raising literature with the statement, "One plant species becomes extinct every day" written across the top of the flier. A young woman glanced at the message and giggled mockingly, "Oh, dear! That's going to keep me awake all night!" Ignorance is bliss.

A single exposure on television or in a newspaper is practically meaningless. So many events happen in the world that a message will be effective only when repeated in many different forms, again and again. One story about whales is not enough. Letters and posters, bumper stickers and rallies, television outcries and lectures are necessary before "saving the whale" becomes a moral stance advocated by a consensus in this country. A lot of work is required by a lot of individuals. This is exactly what is needed to save the plant world from catastrophe. Writing letters to newspapers is a good way to start sending out the general message.

Radio is also a good medium. Vast numbers of people listen to radio talk shows. Learn about the problem and then call radio station talk shows in your area. This is an effective way to direct widespread attention towards plant conservation, particularly among an audience whose primary concerns are not about plants.

Gardens, Museums and Libraries

Make inquiries at local arboreta and botanical gardens about their propagation of endangered species. Ask if they grow rare species. If they do not have such a collection, ask them why not. Even at zoos, which frequently are located in parks, you can inquire about endangered plant species. One of the most famous zoos in the world, the San Diego Zoo, is actively concerned with plant conservation. Make certain that administrators of gardens understand that the public regards plant conservation as an integral part of any important plant collection and that conservation is expected of them. If the garden you visit has a poster about a living plant display area, request a future display on endangered plants. Many garden administrators enjoy interactions with visitors. They frequently will go out of their way to accommodate requests from the public.

Museums, and even public libraries, are good places for displays. A request to a local museum for a display on endangered plants and deforestation will alert museum officials to a public need. You may need to ask your friends to submit similar requests in the weeks following to help reinforce the request. Libraries may be willing to provide space for a conservation display as well.

Spread the Word

Awareness is the first step toward understanding and action. Perhaps the single most important step you can take is to spread the word. Talk to your friends and neighbors. Even some professional botanists are surprisingly almost totally unfamiliar with the plant extinction crisis. If plant scientists are unaware, it should not come as a great shock that John and Mary Doe also are ignorant. Discuss some of the data from this book with your family and friends. Most people just don't realize how plants contribute to their everyday lives. Few are really aware of medicines or other products that plants provide. A few facts about Rauwolfia and hypertension, or Digitalis and heart disease usually catch people's attention.

Be a Joiner

Local, statewide, national, and even international conservation organizations are involved in the efforts to save plants. These groups need your support in terms of membership, audiences, and money. Be sure to earmark your donation for plant conservation work. Even within conservation groups, plants have a tough time competing with animals. Despite the overwhelming numbers of endangered plant species, plants merit only one IUCN Red Data Book; animals groups each receive individual Red Data Books. Many members of conservation groups are very receptive to information about plants and are eager to become involved. Besides general conservation groups, some are devoted to wildflowers, such as California's Native Plants Society, a successful group that publishes an informative journal called *Freemontia.* The journal covers both the natural history of California native plants and conservation issues. Other organizations have similar goals and activities, but they need additional members. Even if you are not an activist, your presence on the membership roster of any conservation group will help lend clout. The bigger the membership, the bigger the stature.

Many private and corporate foundations in this country fund worthy projects but conservation efforts usually are not high on all of their priority lists. Write to these foundations and ask for lists of conservation projects that they currently are funding. Inquire about how they rate these projects.

Beating the Bulldozers

Except for large metropolitan areas where open land is just a memory, many communities still have patches of wild flowers. Polite requests to the mayor's office for local ordinances to protect wild flowers will help bring the matter to public officials' attention. People often will protect rare species if they know these plants exist, but first they must be informed. For smaller cities and towns, being the home of an endangered species brings much civic pride. If such a species is discovered, the local newspaper is usually happy to inform its readers. Definitively knowing whether or not an endangered species is located in your area can be difficult. You could consult local botanists and museums; states often have endangered species lists, too. The tricky part is that not all endangered species are recognized as being in trouble. Botany or ecology faculty at local colleges often will know of surrounding areas where rare plants can be found. If you suspect that one or more of the local wildflowers are truly rare, try to confirm this with local authorities before swinging into action. The plant possibly could be rare only in your area. If so, it may not be worth the effort to save, since another region nearby may have vast expanses of this same supposedly rare plant. Once you are convinced that the species is rare, and the project is worthwhile, the most effective way to gain protection is to persuade city or county governments to set aside small preserves. Because land is so valuable, you may have to accept just official recognition that a rare species exists. When a rare plant occurs on private property, two things should be done: alert the owner, and try to get a small news item in the local newspaper. Media coverage brings public recognition of the landowner's custodianship. This will tend to reinforce the owner's responsibility.

Land development—either clearing fields for buildings or farms, cutting hills for roads, or flooding valleys to create dams—decimates vast numbers of wild species. In some countries (such as South Africa), people concerned about plants are allowed to collect and remove vegetation before it is destroyed. People dig up bulbs and corms, remove cuttings and transplant shrubs. In Mexico, on the other hand, there is not much interest in saving plants that are headed for decimation. Mexico has extensive land-clearing programs for agriculture, where acres of cactus plants are cleared and left in heaps to rot at the edges of roadsides. Exportation to other countries could save some of these plants, but international treaties ban trade in endangered species.

Consumerism

Active consumerism is another way to help the conservation effort. Plant collectors, by collecting and selling wild species, have devastated entire species. Some of the hardest hit plants include many American cactus species, numerous Asian orchids, various carniverous plants, and some small bulbous varieties. Cactus or carniverous plants should not be purchased unless accompanied by evidence that the materials were nursery-grown from seed, not wild-collected. Don't hesitate to ask your nurseryman about the origin of any suspicious plant. He may be unaware that his Venus Flytraps are endangered or that the *Ariocarpus* (living rock cactus) were the result of illegal trade. Enough disapproval from consumers eventually trickles back to the plant collectors and retards trade in these species. Be particularly suspicious of advertised "jungle-collected" plants. Wildflowers, and slipper orchids frequently are advertised in North America. Ask if these are nursery-grown or wild-collected stock and refuse the purchase the latter. Make it clear that you will not support further degradation of the world's plants.

Get Informed

If you want to be taken seriously, you must be informed. It is important to know the facts. How fast is the rate of deforestation? Where is the situation most critical? How many plants are useful? Which states are doing the most to help the problem? Where does one join to help? The answers to these questions and others are in this book but the answers will change with time. Keep up with the changing situation through information from books and magazines in libraries. Ask your local library to order conservation-oriented periodicals. Ask the reference desk to research and then order new titles on conservation. Organize a few friends to request the same kinds of materials from the same libraries. Libraries respond to the wishes of their publics.

In the current period of budget restrictions, your local library may be unable to purchase the titles you want. However, all libraries are part of a network that extends borrowing privileges. Your library can obtain newer titles for you through interlibrary loans.

A Cause for All Ages

Young people make up one of our greatest forces for change. Schools are always looking for guest lecturers. If you and your group are determined to save some plants or generally educate those around you, approach your local high, junior high, or even primary schools. Ask if you may tell the children about the importance of plants and the current predicament in the plant kingdom. Also, ask if the school library subscribes to any conservation-oriented periodicals.

Just as important are the older, retired folks. They are just as good an audience, and usually are even more appreciative. Visit and speak with them. Ask them to get involved. What a joy it could be for a senior citizen to suddenly find an important way to spend his or her leisure time *and* to make a valuable contribution to maintaining the delicate ecological equilibrium of our planet. Who could pass on a better legacy?

We all occasionally feel that we are up against helpless or hopeless situations. The truth is that many individuals working for the same cause can't move a mountain—but they *can* save some plants.

Check List

Individuals or groups may find this checklist helpful in their attempts to do something about the plant problem.

1) Write to elected representatives.
2) Talk to non-gardening friends.
3) Work with local and national conservation groups.
4) Apply pressure to local gardens.
5) Write to newspapers and magazines.
6) Discuss the problem on radio talk shows.
7) Join and send money to conservation groups and earmark the money for plant conservation.

8) Ask plant societies to set up gene banks.
9) Ask foundations for details on conservation projects.
10) Ask local officials about local ordinances to protect plants.
11) When you buy succulents or orchids, ask for notes on breeding. Patronize nurseries that breed stock.
12) Retrieve rare species from vacant land before developers arrive.
13) Ask museums to hold displays on species extinctions and conservation.
14) Speak to groups of school children and/or older, retired people.
15) Ask libraries to obtain books and periodicals on plant conservation.

Campanula portenschlagiana

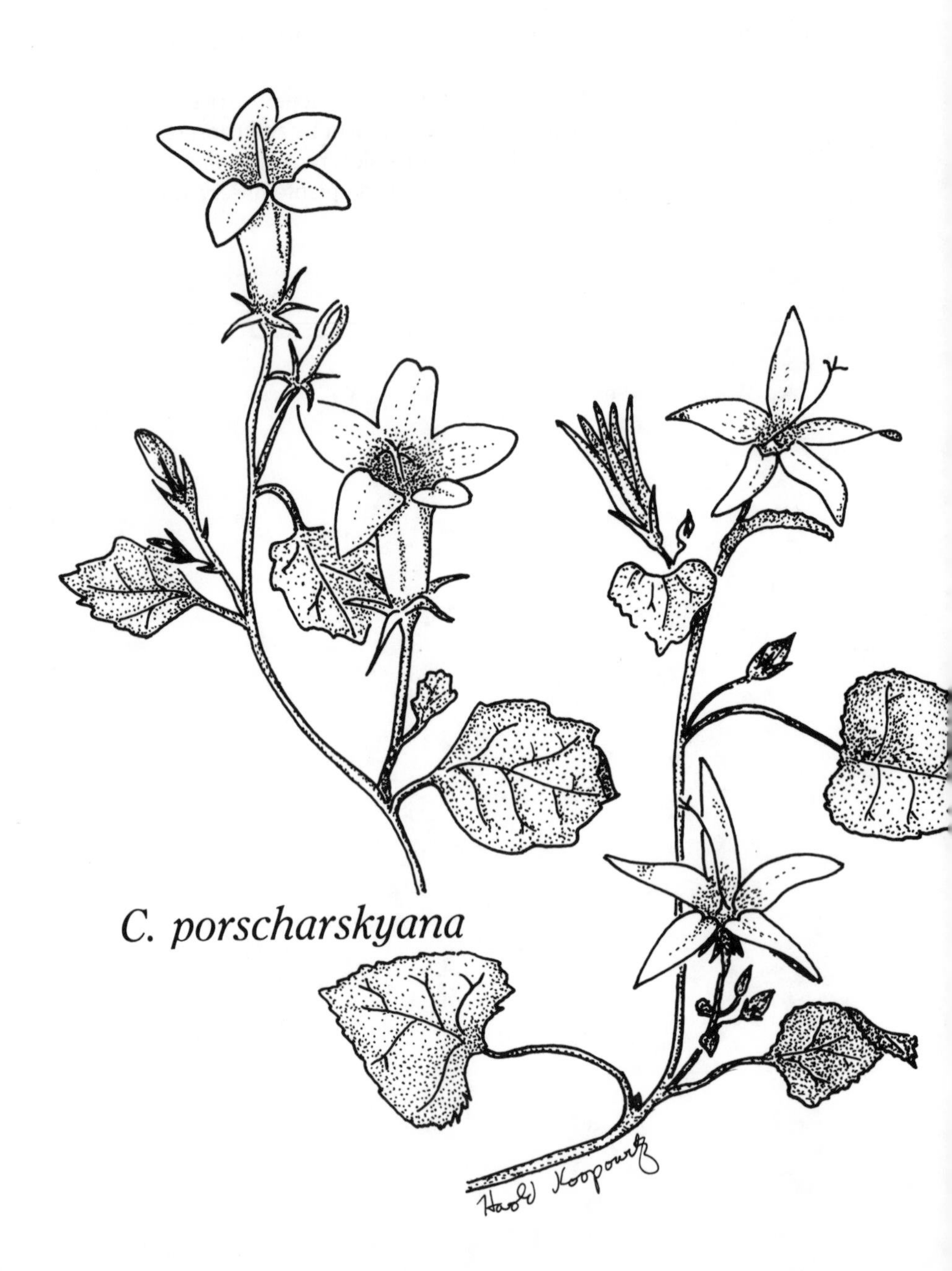

Case History: The Bellflowers

Many of our case histories have been dramatic tales about unusual plants or trees. Most species simply are not dramatic. They live ordinary lives and often are neither beautiful nor very showy. These plants, the simple plants, are the ones that make up the majority of names on the Red Data lists. They receive little attention, and are the least likely to be saved. The group of plants called bellflowers contains many of these species.

One of the greatest English writers about plants was Reginald Farrer who traveled to the mountain ranges of Europe and Asia in search of botanical treasures. He wrote many books about his travels and the plants he encountered. "The English Rock Garden" will remain a classic in the annals of horticulture. He wrote more than fifty pages describing many of the bellflowers that belong to the genus *Campanula*. One of these was a little blue bellflower with one of the longest scientific names: *Campanula portenschlagiana.* He described the plant as being plentiful and common in its native Yugoslavia, ". . .abounding through eastern Dalmatia, in all the walls and fissures—as in the cliffs by Almissa and Macarskea and throughout the province." Quoting Farrer further, "True it is that we might have hoped good things of those smooth and hairless leaves, but who could ever have foretold the generosity of their abounding masses and the way they are hidden beneath the inordinate profusion of starry violet cups?" These words were written in 1913. Seventy years later, the small bellflower is considered so rare in Yugoslavia that it has been incorporated into the list of European endangered flora circulated to botanic gardens. *C. portenschlagiana* seems to be safe in cultivation since it is popular in gardens. The species demands little attention and is always generous with its late summer sprays of lilac blue bells.

Another bellflower species from the Yugoslavic mountains has a name almost as cumbersome: *Campanula porscharskyana.* This plant grows a neat tufted cushion of soft green heart-shaped leaves just like its cousin, but this species becomes undisciplined and produces trailing flowering stems with flat blue stars. This bellflower is now rare, too, in its native habitat in the Dalmatian Mountains. Fortunately, the popularity of these two species in rock gardens and window boxes should keep them from completely disappearing.

Like so many other small plants, the reasons for the demise of these bellflowers are rather obscure. As far as we can tell, they were not subjected to sudden or violent destruction in their native habitats. Gradual attrition seems to have eaten away at their numbers. With species of this sort, their decline is seldom noticed until it is too late.

CHAPTER 20

A Prognosis

When we first decided to write this book, very little attention was being given to plants, even though more plant species became extinct in a single year than animals had in all of the preceeding centuries. We are finally seeing a growing awareness of the plant problem. The World Wildlife Fund now speaks about animals and plants in the same breath. The world's oldest and most prestigious conservation group recently added plants to its name and is now known as the "Fauna and Flora Preservation Society." All over the world, animal conservationists are realizing that animal and plant conservation are intertwined concerns. One can't be done without the other. For every plant species that disappears, perhaps as many as eleven animal species will follow.

While this growing awareness is taking place on a global scale, time is running out. Our bureaucracies move at a snail's pace while the problem escalates at the pace of a jackrabbit. Even organizations that are truly concerned about the problem, such as the IUCN and FAO, seem to generate an overwhelming number of acronyms and small bureaucracies. Plant conservationists, like mankind, seem to have mastered the art of making resolutions. The difficult part is to turn resolutions into accomplished tasks.

How simple it is to point a finger of blame when a problem exists. In this case, however, it's not so easy. Until about twenty years ago, when the alarm first went out that all was not well within the plant kingdom, the villain was ignorance. Very few people had a clear idea that the isolated destruction of habitats was wreaking such havoc on the ecosystems of the world. Since the 1960s, more factors have entered the picture. The devastating poverty in developing nations, where financial realities outweigh the longterm hazards, is a major obstacle. The forests in many of these developing countries are plundered, and only wealthy agriculturists or outside investors benefit from the deforestation. For the most part, these people know they are ruining the country's resources but they are motivated by greed, not conscience. Finally, apathy is to blame through much of the Western world. Many people have heard vague mutterings about an extinction crisis but since the victims are plants, not animals, they have tuned out and haven't even heard that animals and humans will similarly suffer.

As long as we're pointing fingers, plant scientists, as a group, must shoulder some of the burden. In the United States during the past two decades, only a few concerned scientists, such as Peter Raven, Hugh Iltis, and Edward Ayensu, have spoken out on these issues. The community of plant scientists has been curiously quiet. The people speaking out were too few in number and not visible enough to make an indelible impression on the public. In today's electronic era, any message that is not repeated time and time again through television, radio, newspapers, and magazines usually is not really digested. People respond only after they have heard the same message over and over. Plant scientists apparently were too busy in their laboratories to help sound the alarm. For the most part, they professed disinterest in the whole matter. If they had banded together and spoken in a clear, unified voice of the need for immediate plant conservation measures, perhaps the situation would not be as grave as it is now. The scientists certainly had the stature to carry off a worldwide plea for help but, it seems, not the concern. Many more plant scientists have recently become interested in the problem, yet it is almost too late.

Part of the tragedy is that even though general concern is beginning to surface, the world's current financial situation is bleak. We are gripped in a global monetary crisis with entire nations either defaulting on their loans or barely surviving with their payments for oil. As this is being written, budgets for work on endangered species are being cut in the United States. Much of the world cannot afford the high cost of developing and managing preserves for endangered plants. Rich nations that could possibly afford such activities seem to be spending most of their budgets on armaments. Not much is left to save the world's threatened organisms.

Looking at mankind objectively, our concept of the universe has barely changed through the centuries. We know our world is not flat, but we still live in a kind of fantasy world. While we realize the the earth is not the center of the universe but rather a mere pebble circling a not-very-distinguished star, we still conduct our business as if humans were the only significant entities anywhere. People refer to the dark, "animal" side of human nature, but perhaps this perspective is wrong. The truth is, most animals live in harmony with their ecosystems. They may kill other organisms but only for survival, not for greed or fun. Perhaps it is the "human" side of man that is responsible for man's disreputable activities—poisoning the environment and haphazardly destroying organisms that share our habitats. In many respects humanity can be likened to a disease out of control. The germs are multiplying furiously. If they do not come to terms with their host, they will kill both the host and themselves. We seem to be doing just that.

We are admittedly offering a gloomy prognosis. But humans traditionally tend to respond to immediate crises, not projected crises. Longterm problems are rarely solved; they tend to languish and turn into short-term problems. We do believe that some plant species will be saved, but we suspect that the percentage will be rather small. We predict that the world will see a variety of stop-gap efforts in the years ahead; in most cases, they will come too late.

This is our prediction, but we sincerely hope we're proven wrong. Perhaps through a combination of greater awareness in developing countries, an increased number of gene banks, and various last-gasp efforts throughout the world, we will save a fair number of our plants. No one knows just how many can be saved. The numbers depend on how many people decide to take action and how soon they begin to act. Certainly we can't save the majority of our endangered species. It's too late for that. But whatever the number turns out to be, we should keep in mind that half a loaf is better than none at all.

VI.

APPENDICES

CONSERVATION ORGANIZATIONS

Some of the organizations listed here do not deal primarily with plant conservation, but they are sympathetic to the cause. Either directly or indirectly, they do contribute towards plant conservation.

United States

American Horticultural Society, Inc.
Mount Vernon, Virginia 22121

Association of Western Native Plant Societies
℅ Anne Kowalishen
4949 N.E. 34th
Portland, Oregon 97211

Biological Institute of Tropical America
P.O. Box 2585
Menlo Park, California 94025

California Native Plant Society
2380-D Ellsworth
Berkeley, California 94704

Defenders of Wildlife
1244 19th Street, N.W.
Washington, D.C. 20036

Growers Network
℅ Kent Whealy
RFD 2
Princeton, Missouri 64673

National Audubon Society
950 3rd Ave.
New York, New York 10022

National Parks and Conservation Association
1701 18th Street, N.W.
Washington, D.C. 20009

National Wildlife Federation
1412 16th Street, N.W.
Washington, D.C. 20036

Natural Resources Defense Council
Attn: Faith Campbell
1725 I Street, N.W.
Washington, D.C. 20006

The Nature Conservancy
1800 N. Kent Street
Suite 800
Arlington, Virginia 22209

New England Wildflower Society
Garden in the Woods
Hemenway Road
Framingham, Massachusetts 01701

The Rare and Endangered Native Plant Exchange
℅ New York Botanical Garden
Bronx, New York 10458

Sierra Club
530 Bush Street
San Francisco, California 94104

The Wilderness Society
1901 Pennsylvania Ave., N.W.
Washington, D.C. 20006

World Wildlife Fund—U.S.
910 17th St., N.W.
Suite 619
Washington, D.C. 20009

Great Britain

The Fauna and Flora Preservation Society
% The Zoological Society of London
Regent's Park
London NW1 4RY

The Garden History Society
Secretary, Mrs. M. Batey
12 Charlburg Road
Oxford, OX2 6UT

International Dendrology Society
Secretary, Mrs. M. Eustace
Whistley Green Farmhouse
Hurst, Reading
Berkshire RG10 0DU

The International Organization for Succulent Plant Study
% D. R. Hunt
Royal Botanic Gardens
Kew, Richmond
Surrey

Netherlands

The International Society for Horticultural Science
Ministry of Agriculture
Bezuidenhoutseweg 73
The Hague

Switzerland

International Union for the Conservation of Nature and Natural Resources
IUCN/Secretariat
Avenue du Mont-Blanc
1196 Gland-Suisse

RED DATA LISTS OF PLANTS

Here are some Red Data Lists of some threatened plants. This list is by no means complete. Note the obvious lack of lists from tropical countries.

AFRICA

Hall, A.V. 1982. Rare plants gazette. Bolus Herbarium, Rondebosch. 30p.

Hall, A.V., de Winter, M., de Winter, B. and S.A.M. oosterhout. 1980. Threatened plants of Southern Africa. South African National Scientific Programmes Report no.45. Cooperative Sciences Programmes, Council for Scientific and Industrial Research, Pretoria. 244p.

ASIA

Akademiya Nauk Arm SSR Botanicheskii Institut. 1979. List of rare and disappearing species of the flora of Armenia. Erevan. 27p. (in Russian and Armenian).

Bijaschev, G.S. and M.S. Bajtenov. 1981. Red data book of Kazakh SSR. Rare and endangered species of animals and plants. Part 2. Plants. Alma-ata. 263p. (in Russian).

Borodin, A.M. et al. 1978. Red data book of USSR. Lesnaya Promyshlenost Publishers, Moscow. 459p. (in Russian).

Charkevitch, S.s. and N.n. Katchura. 1981. Rare plant species of the Soviet Far East. Moscow. 232p. (in Russian).

Chopik, V.I. 1978. Rare and threatened plant species in the Ukraine. Kiev. 211p. (in Russian).

Jain, S.K. and A.R.K. Sastry. 1980. Threatened plants of India: a state of the art report. Botanical survey of India, Indian Botanic Garden, Howrah-711 103. 48p.

Kononov, V.N. and G.A. Shabanova. 1978. The rare and endangered plants of Moldavia. Kishinev. 27p. (in Russian).

Malyshev, V.L. and K.A. Soboleuska. 1980. Rare and endangered plant species of Siberia. Hayka. 223p. (in Russian).

Malyshev, L.I. and G.A. Peshkova. 1979. They stand in need of conservation: rare and endangered plants of Central Siberia. Nauka, Siberian branch, Novosibirsk. 174p. (in Russian).

EUROPE

Aymonin, G.G. 1977. Studies on the loss of plant species in France. Report no. 3. General list of species requiring protective measures. Société Botanique de France, Direction de la Protection de la Nature, Ministère de la Culture et de l'Environnement. Paris. 58p. (in French).

Council of Europe. 1977. List of rare, threatened and endemic plants in Europe. Kew, Surrey. Nature and Environment Series, no. 14. 286p.

Holub, J., Prochazka, F. and J. Cernovsky. 1979. List of extinct, endemic and threatened taxa of vascular plants of the flora of the Czech Socialist Republic (first draft). Presalia, Praha, 511: 213-7.

IUCN Threatened Plants Committee. 1977. List of rare, threatened and endemic plants in Europe. Nature and Environment Series no. 14. Strasbourg.

Lojtnant, B. and E. Worsoe. 1977. Preliminary list of the status of the Danish flora. Aarhus, Copenhagen. (in Danish).

Nordic Council of Ministers. 1978. Threatened animals and plants in the Nordic countries. Report NU.A 1978:9 of the Nordic Council of Ministers. Stockholm. (various Scandanavian languages).

Perring, F.H. and L. Farrel. 1977. British red data books:1. Vascular plants. Society for the promotion of Nature Conservation. Nettleham. Linc. 124p.

Rauschert, S. et al. 1979. Lists of extinct and threatened ferns and flowering plants in the German Democratic Republic. Kulturbund der DDR. Berlin. (in German).

Sainz Ollero, H. and J.E. Hernandez-Bermejo. 1981. List of the endemic dicotylendons of the Iberian peninsula and Balearic Isles. Madrid. 111p. (in Spanish).

NORTH AMERICA

Ayensu, E.S. and R.A. DePhillips. 1978. Endangered and threatened plants of the United States. Smithsonian Institute. Washington, D.C. 403p.

Fairbrothers, D.E. and M.Y. Hough. 1975. Rare or endangered vascular plants of New Jersey. New Jersey State Museum Scientific Notes 14: 1-53.

Henderson, D.M. et al. 1977. Endangered and threatened plants of Idaho. A summary of current knowledge. Univ. of Idaho. Idaho Natural Areas Council. 72p.

Hodgdon, A.R. 1973. Endangered plants of New Hampshire: a selected list of endangered species. Forest Notes 114: 2-6.

Kershaw, L.J. and J.K. Morton. 1976. Rare and potentially endangered species in the Canadian flora—a preliminary list of vascular plants. Can. Bot. Assoc. Bull. 9(2): 26-30.

Maher,, R.V. et al. 1979. The rare vascular plants of Saskatchewan. Syllogeus no.20. National Museums of Canada, Ottawa.

Nelson, B.B. and R.E. Arndt. 1980. Eastern States endangered plants. Bureau of Land Management. U.S. Department of the Interior. 109p.

Riskin, D. 1975. Provisional list of Texas' threatened and endangered plant species. Texas Organization of Threatened and Endangered Species. Texas Parks and Wildlife Department.

Stauffer, M.R. 1975. Inventory of rare and endangered plants of New York. Proc. Roch. Acad. Sci. 12(4): 400.

PACIFIC

Given, D.R. 1981. Rare and endangered plants of New Zealand. A.H. and A.W. Reed Ltd., Wellington. 154p.

Hartley, W. and J. Leigh. 1979. Plants at risk in Australia. Occasional Paper no. 3. Australian National Parks and Wildlife Service. Canberra 80p.

Leigh, J. and R. Boden. 1979. Australian flora in the endangered species convention CITES. Australian National Parks and Wildlife Service Special Publication no. 3. 93p.

Rye, B.L. and S.D. Hopper. 1981. A guide to the gazetted rare flora of Western Australia. Department of Fisheries and Wildlife, Perth.

REFERENCES

This reading list is only a partial selection of materials. Some of the references will expand on some of the data that have been discussed in this book. But beware—the situation changes rapidly and information is quickly outdated.

Anonymous (Ed.). 1977. *National Weeds Conference of South Africa,* 2nd. A.A. Balkema, Cape Town. 344p.

Ehrlich, P. and A. Ehrlich, 1981 *Extinction*. Random House, New York, 305 p.

Frankel, O.H. and E. Bennet. 1970. *Genetic Resources in Plants—Their Exploration and Conservation.* IBP Handbook no. 11. F.A. Davis, Philadelphia. 554p.

Frankel, O.H. and J.G. Hawkes. 1975. *Crop Genetic Resources for Today and Tomorrow.* Cambridge University Press, Cambridge. 492p.

Frankel, O.H. and M.E. Soule. 1981. *Conservation and Evolution.* Cambridge University Press. Cambridge. 327pp.

Hawkes, J.G. and W. Lange. 1973. *European and Regional Gene Banks.* Proceedings of a conference of Eucarpia, Izmir, Turkey 1972. Wageningen, Netherlands.

Horsfall, J.G. 1972. *Genetic Vulnerability of Major Crops.* National Academy of Sciences, Washington, D.C. 307p.

Iltis, H.H. 1978. *Extinction or Preservation: What Biological Future for the South American Tropics.* University of Wisconsin, Madison. 218p.

Lucas, G. and H. Synge. 1978. *The IUCN Plant Red Data Book.* The Gresham Press, Old Working, Surrey.

Miasek, M.A. and C.R. Long. 1978. *Endangered Plant Species of the World and Their Endangered Habitats: a Compilation of the Literature.* New York Botanical Garden. Bronx. 46p.

Myers, N. 1979. *The Sinking Ark.* Pergamon Press, Oxford.

Myers. N. 1980. *The Conversion of Tropical Moist Forests.* National Academy of Sciences, Washington, D.C. 205p.

Prance, G.T. and T.S. Elias. 1977. *Extinction is Forever.* New York Botanical Garden, Bronx. 437p.

Sherbrooke, W.C. and P. Paylore. 1973. *World Desertification: Cause and Effect. A Literature Review and Annotated Bibliography.* University of Arizona, Tucson, 168p.

Simmons, J.B. 1976. *Conservation of Threatened Plants.* Plenum Press, New York. 366p.

Smithsonian Institute. 1975. *Report on Endangered and Threatened Plant Species of the United States.* Committee on Merchant Marine and Fisheries, Serial no. 94-A. House Document no. 94-51. U.S. Govt. Printing Office, Washington, D.C. 200p.

Soule, M.E. and B.A. Wilcox. 1980. *Conservation Biology: an Evolutionary Perspective.* Sinauer Associates, Sunderland.

Specht, R.L. 1974. *Conservation of Major Plant Communities in Australia and Papua, New Guinea.* Commonwealth Scientific and Industrial Research Organization, East Melbourne. 667p.

Synge, H. and H. Townsend. 1979. *Survival or Extinction.* Bentham-Moxon Trust. Kew. 250p.

PERIODICALS

Biological Conservation—Technical papers on animal and plant conservation. Some ecological papers.

Environmental Conservation—A mix of both technical and non-techhnical papers. A must for keeping up with conservation and its attendent problems on a world-wide basis.

Fremontia—Published by the Californian Native Plant Society. Non-technical and contains material on both conservation and appreciation of native plants.

National Wildlife—Primarily on animals, but there are the occasional papers on plants. Non-technical. Published by the National Wildlife Federation, Washington, D.C.

Nature Conservancy News—Has the occasional feature on plants. Non-technical.

Oryx—Published by the Fauna and Flora Preservation Society.

Threatened Plant Committee Newsletter—Very informative and an excellent source of pertinent information. Distribution is primarily to institutions.

INDEX

Other Stone Wall Press books for Conservationists:

These are the Endangered by Charles L. Cadieux.
illustrations by Bob Hines
240 pages, illustrated, 6″ × 9″, hardcover, $16.95

"A must for anyone concerned with identifying the myriad threats to our wildlife and with gaining a better understanding of the issues involved."
National Parks

"A useful reference for students of the endangered species movement in this country."
Prof. James Hardin,
University of Wisconsin

Vanishing Fishes of North America by Dr. R. Dana Ono,
Dr. James D. Williams and Anne Wagner.
Color illustrations by Aleta Pahl.
272 pages, color illustrations, 7″ × 10″, hardcover, $27.50

The plight and importance of over fifty species in North America is chronicled with beautiful paintings and photographs. Many fish species are threatened. Fourteen entire species became extinct in the last century. We can help preserve these unique species by understanding and accommodating their needs. Available May, 1983.

Backwoods Ethics:
Environmental Concerns for Hikers and Campers
by Laura and Guy Waterman.
192 pages, illustrations, 6″ × 9″, paperback $7.95

Noted outdoor magazine columnists shed light on sensitive environmental issues with neighborly warmth and humor to maintain the spirit of wildness. Endorsed by the American Hiking Society.

"The best statement of the concerns of responsible hikers and outdoorsmen of today. It's also fun to read."
Potomac Appalachian

Practical Outdoor Books from Stone Wall Press:

MOVIN' OUT: Equipment & Techniques for Hikers by Harry Roberts.
156 pages, illustrations, index, 6″ × 9″, paperback, $7.95

A thorough introduction to backpacking by an outfitter and wilderness expert. No-nonsense wisdom is accompanied by practical money-saving advice for choosing the best clothing and equipment for your purposes—what you need and don't need. Also, how to best use that equipment you just bought.

"An excellent, down-to-earth book on backpacking information."
International Backpackers Association

MOVIN' ON: Equipment & Technique for Winter Hikers by Harry Roberts.
144 pages, illustrations, 6″ × 9″, paperback, $7.95

Roberts takes his what-works approach to the special problems and techniques involved in winter hiking and camping. Winter backpacking, cross-country skiing, or snowshoeing are all covered.

". . . Emphasizes technique which is all-important to survive in winter . . . a superb book, even if you are just thinking about maybe *going winter camping."*
Backpacker Magazine

Keeping Warm and Dry by Harry Roberts.
144 pages, illustrations, paperback, $7.95

Don't get caught unprepared on your next outing! Sudden cold or drenching rain often hit the careless. Let Harry tell you how he would prepare for a camping, fishing, or hunting trip in any part of the country. Solid what-works advice, field-tested techniques and no-nonsense looks at today's fibers and gear.

Introducing Your Kids to the Outdoors by Joan Dorsey.
Revised edition, 144 pages, illustrations, photographs, appendices, 5½″ × 8½″, paperback, $8.95

For the outdoors people who want to become outdoors families. Taking the kids with you fishing, backpacking, cross-country skiing, and bicycling involves special preparation—but the rewards are great! See the wilderness with your children, talk without distractions, and share responsibilities. Dorsey carefully discusses what adults must plan for: Food, safety on the trail, equipment that both fits the infant or small child and serves a purpose.

"Highly recommended for its thoroughly sensible, encouraging advice . . ."
Blair & Ketchum's COUNTRY JOURNAL

Enjoying the Active Life After Fifty by Ralph Hopp.
Foreword by Arthur S. Leon, M.D.
192 pages, photographs, 6″ × 9″, paperback, $7.95

An exercise enthusiast provides older readers basic, practical information for enjoying a wide variety of aerobic sports and recreational activities. Eighteen activities and accompanying illustrations are covered with complete bibliographies of books and magazines.

"Activity is the key to health and staying younger. This book is testament to the fact that its philosophy works."
San Francisco Examiner

Backpacking for Trout by Bill Cairns.
Introduction by Lefty Kreh
200 pages, photographs, 6″ × 9″, hardcover, $16.95

The former technical director of the Orvis Company discusses backpacking equipment and technique, as well as an excellent synopsis of the essentials of fly and spin fishing. The only book to get anglers into serene, more productive waters to catch more trout. Excerpted and used by Trout Unlimited.

"The most complete guide to trout fishing and backpacking available."
Massachusetts Audubon Society

The Natural World Cookbook—Complete Gourmet Meals from Wild Edibles, by Joe Freitus
301 pages, illustrations, index, 7″ × 10¼″, hardcover, $25.00

Complete and comprehensive, this is a twenty-year collection of proven gourmet recipes for wild plants, fish, fowl, and game. Hundreds of recipes are carefully presented with clear how-to-find illustrations by Salli Haberman. Freitus has taken up where Euell Gibbons left off.

"Everything for the wild foods connoisseur!"

Scott Wildlife News

Wild Preserves—Illustrated Recipes for Over 100 Natural Jams and Jellies by Joe Freitus
192 pages, illustrations, 6″ × 9″, paperback, $7.95

Preserving wild fruit is a simple and delicious job, but the canner must know about the fruit and its preservability before starting. Freitus discusses pectin levels, sugar content and cooking procedures.

"A happy combination of plant identification information and recipes makes this book perfect for the pockets of hunters, camera fans, hikers, followers of Euell Gibbons or just weekend wanderers. Joe Freitus has written the creative canner's bible!"

The Conservationist

Ask for these books at your bookstore, or send your check for the total amount plus $2.00 for shipping and handling to

STONE WALL PRESS, INC.
1241 30th Street, N.W.
Washington, D.C. 20007

